PENGUIN BOOKS

# THE GLOBAL WARMING READER

Bill McKibben is the author of a dozen books about the environment, including *The End of Nature*, the first book for a general audience about global warming. He is a founder of the grassroots climate campaign 350.org and is a frequent contributor to such publications as *The New York Times*, *The Atlantic*, *Harper's Magazine*, *Mother Jones*, *The New York Review of Books*, *Granta*, *Rolling Stone*, and *Outside*. He is a fellow of the American Academy of Arts and Sciences and has been awarded Guggenheim and Lyndhurst fellowships, the Lannan Prize, and honorary degrees from a dozen colleges. A scholar in residence at Middlebury College, he lives in Vermont with his wife, the writer Sue Halpern, and their daughter.

# THE GLOBAL WARMING READER

## A Century of Writing About Climate Change

EDITED WITH AN INTRODUCTION BY

## BILL McKIBBEN

Penguin Books

PENGUIN BOOKS

Published by the Penguin Group

Penguin Group (USA) Inc., 375 Hudson Street, New York, New York 10014, U.S.A.
Penguin Group (Canada), 90 Eglinton Avenue East, Suite 700, Toronto,
Ontario, Canada M4P 2Y3 (a division of Pearson Penguin Canada Inc.)
Penguin Books Ltd, 80 Strand, London WC2R 0RL, England
Penguin Ireland, 25 St Stephen's Green, Dublin 2, Ireland (a division of Penguin Books Ltd)
Penguin Group (Australia), 250 Camberwell Road, Camberwell,
Victoria 3124, Australia (a division of Pearson Australia Group Pty Ltd)
Penguin Books India Pvt Ltd, 11 Community Centre,
Panchsheel Park, New Delhi – 110 017, India
Penguin Group (NZ), 67 Apollo Drive, Rosedale, Auckland 0632,
New Zealand (a division of Pearson New Zealand Ltd)
Penguin Books (South Africa) (Pty) Ltd, 24 Sturdee Avenue,
Rosebank, Johannesburg 2196, South Africa

Penguin Books Ltd, Registered Offices:
80 Strand, London WC2R 0RL, England

First published in the United States of America by OR Books 2011
Published in Penguin Books 2012

1  3  5  7  9  10  8  6  4  2

Pages 423 and 425 are an extension of this copyright page

Anthology selection, introduction and introductory comments
to each essay copyright © Bill McKibben, 2011
All rights reserved

ISBN 978-0-14-312189-3
CIP data available

Printed in the United States of America

# CONTENTS

## PART II: POLITICS

# Introduction

I write these words in May of 2011, the week after a huge outbreak of tornadoes killed hundreds across the American South; it was the second recent wave of twisters of unprecedented size and intensity. In Texas, a drought worse than the Dust Bowl has set huge parts of the state ablaze. Meanwhile, the Army Corps of Engineers is moving explosives into place to blow up a levee along the Mississippi River, swollen by the the third "100-year-flood" in the last twenty years—though as the director of the Federal Emergency Management Administration noted at the end of 2010, "the term '100-year event' really lost its meaning this year." That's because 2010 was the warmest year recorded, a year when 19 nations set new all-time high temperature records. The Arctic melted apace; Russia suffered a heat wave so epic that the Kremlin stopped all grain exports to the rest of the world; and nations from Australia to Pakistan suffered flooding so astonishing that by year's end the world's biggest insurance company, Munich Re, issued this statement: "The only plausible explanation for the rise in weather-related catastrophes is climate change. The view that weather extremes are more frequent and intense due to global warming coincides with the current state of scientific knowledge."

And that's not the bad news. The bad news is that on April 6, the U.S. House of Representatives was presented with the following resolution:

> "Congress accepts the scientific findings of the Environmental Protection Agency that climate change is occurring, is caused largely by human activities, and poses significant risks for public health and welfare."

The final vote on the resolution? 184 in favor, 240 against. When some future Gibbon limns the decline and fall of our particular civilization, this may be one of the moments he cites.

In some sense, this anthology is an attempt to deal with the paradox that, though everyone (scientists, the insurance industry) we depend upon to analyze risk tells us we are facing the gravest crisis in human history, our political system here and around the world is doing nothing. The Democratic-controlled Senate last summer refused even to take a vote on the mild, modest, even tepid climate change bill that came before it; President Obama has wasted no political capital pushing for such legislation. There are many obstacles to change, of course: the inertia that comes with our dependence on fossil fuels; the powerful vested interests of the most profitable industry the world has ever seen. But there's also a fundamental lack of understanding about how high the stakes are, about what we can do, and most of all about what it means to be watching the only world humans have ever known change in deep and dramatic ways. We can't just be stunned—that seems to lead to denial, to inaction. We need to feel what's happening, not just in our overheating bodies but in our minds and spirits too.

Perhaps the first thing to realize about global warming is what a new idea it is. The first few papers in the Science section of this anthology may belie that notion, but it is important—while giving great honor to Arrhenius and the few others who first conceived the idea—to realize that not even other scientists were paying attention until very recently. It was not until the late 1980s that climate change broke into the open: if you had to assign a date, it would be the June afternoon when NASA's James Hansen testified before a House committee, a story that ran above the fold in the *New York Times* the next day and signaled the start of a pervasive new worry for human beings.

For some while, the scale of that worry was not clear. Those of us who thought it was likely to be catastrophic (I wrote the first book for a general audience about this topic, in 1989, and it bore the fairly definitive title *The End of Nature*) had to wait a few years for science to conclusively confirm those fears. The world's climatologists cobbled together what may turn out to be one of the most important organizations in the planet's history: the Intergovernmental Panel on Climate Change (IPCC), which by 1995 had concluded that humans were warming the planet and that it was likely to be serious.

In the fifteen years since, it's as if the planet itself has been conducting a rigorous peer review of that science, just to make sure it was right. We've had a succession of incredibly warm years—already, in fact, the planet has warmed on average about one degree Celsius. And it turns out that's enough to cause huge physical changes, which are occurring faster than (habitually conservative) scientists would have predicted even a few years ago. For instance:

- The Arctic has lost twenty-five percent of its summer ice cover, and the melt on Greenland is proceeding with unnerving speed.

- The hydrological cycle is fundamentally disrupted, with both more droughts and far more extreme rainfall events.
- The oceans have become steadily more acidic as seawater absorbs carbon dioxide, changing its chemistry.
- Forest fires are raging with newfound intensity, and forests in the boreal regions are dying from onslaughts of insects once kept in check by cold winter temperatures.

In other words, virtually every major physical system on the planet is now in a state of remarkable flux. In the geological blink of a blink of a blink of an eye, we've managed to bring the epoch that scientists call the Holocene to an end. Since that 10,000-year period of climatic stability marked the rise of human civilization, its conclusion should worry us. And, indeed, around the world we can already see the widespread effects of this warmer, wilder weather on human life:

- Crop yields have become erratic, with serious busts as heat waves wipe out whole growing regions.
- Mosquito-borne diseases, for example dengue fever, have spread rapidly, and have now reappeared even in North America.
- Political tensions have begun to flare over water shortages and refugee fears: India, for instance, has constructed a long wall on its boundary with Bangladesh, a country exquisitely vulnerable to global warming.
- Small, low-lying islands have been evacuated as rising seas have made habitation impossible.

In the Science section of this book, I've included some of the basic documents that allow us to understand just what's going on. Some of them are historic—it's important that we remember how we learned of this problem and how our understanding has evolved. Some are current. But all should be, at some deep level, unsettling. If they have a basic message, it's this: we've taken the stability of our planet entirely for granted. That stability has let us largely ignore the physical world—it's been the backdrop for our political and economic lives. But as it happens, that stability was precarious, and it has now been upset. We were born on one world (a world we could count on to support us) and now we live on a new world, without any such assurance.

But the science of global warming is, in some sense, the easy part (though the difficulty of modeling the impacts of carbon, and of measuring its many effects, should never be underestimated—these researchers are real heroes). What's been harder, almost from the start, is the politics. To do anything about global warming would require doing something about fossil fuel—in particular, it would require us to stop burning it. Which is only about the hardest task you can imagine.

For one thing, fossil fuel is what we do. If an alien landed here, he could be forgiven for describing us as flesh-colored appliances for combusting coal, gas, and oil. So getting off it is not like the other environmental challenges we've faced; it doesn't require a small technical fix à la catalytic converters on exhaust pipes. It's not so much that some mistake is being made that needs fixing; it's not like when BP manages to stab a hole in the bottom of the ocean that they can't patch. It's what happens all day every day. A coal-fired power plant doesn't need an accident to wreck the planet; it performs that task constantly. And of course, on some smaller level, each of us does the same.

It's also worth remembering just how much profit the fossil fuel system represents. ExxonMobil made more money in each

of the last three years than any company in the history of...
money. In our political system, that extraordinary profitability
allows it and other energy companies an almost infinite ability
to wield influence, especially when all they must do to win is
delay action. Delay is simple to arrange—this past summer, for
the twenty-first straight year of the global warming era, the U.S.
Congress decided not to do anything about the problem. "I hate
to tell you, I just don't wake up thinking about it," said Tennessee
Republican Bob Corker, who promised to vote against the bill.

It's true that this partisan craziness is strongest in the United
States. But it's also true that there's another challenge, at least
as big, to political progress on climate. That's the position of the
developing nations, who look around and notice that the nations
that have already developed did so by burning big piles of coal
and big pools of oil. For China and India, and then for everyone
else lined up behind them, that's the easiest way to proceed as
well, and so it's the path down which they've rushed in the last
decades: by now China has passed the U.S. as the largest emitter
of $CO_2$, almost entirely due to coal-fired power. (Of course, it's
always worth bearing in mind the country's giant population—in
per capita terms they're still pikers compared to Americans.)
The West has no moral right to tell China to choose some other
path—but there's no practical alternative. And so the Chinese
are caught: full steam ahead and lose the Himalayan ice, or slow
down and lose the record economic growth that keeps their society stable.

The only way out of that box is if the West can figure out
how to share some of the wealth it's built up in two centuries of
carbon-burning with the rest of the planet, preferably in the form
of renewable energy technology. It would take a full-on commitment—it would take a lot of money. A hundred billion dollars

annually is the figure diplomats have been tossing around, but earlier this year America's climate negotiators said the money would have to come from "private sources," not taxes. Because they're scared of the politics. The politics are, probably, impossible—unless somehow we can build a movement that can really push.

That movement can only come when we feel, deep down, the impact of what's happening around us. The third section of this book may be the most important, the place where writers have the most to add. So far, the literature of global warming remains fairly thin—and the reason, I think, is that the magnitude of what we're doing is so hard to take in. For all of our civilization, we've thought mostly about the relationship between human beings and human beings. That's where the conflict and the action have been. But that's now changing, and changing fast. The planet itself is becoming a key character.

If we could feel in our bones just exactly what it is that's happening—if we really understood that the decisions we make in the next Congress or Parliament or Politburo will determine what happens for essentially the entire human future—we could maybe spur ourselves to action. That's a difficult conceptual leap; our genes haven't really equipped us for thinking on this scale. So this will be like an exam, and maybe a final exam, for how well our culture has readied us to see beyond those instincts. Yes, we need new wind turbines and solar panels. But really, most of all, we need new metaphors. And we have very little time to find them.

*— Bill McKibben*
*Middlebury, Vermont*
*May 2011*

# PART I
# SCIENCE

*Svante Arrhenius was one of those polymaths rarely encountered in the modern, specialized era. A Swede who showed a prodigious early talent for math, he won the Nobel Prize in Chemistry in 1903 for his work on the conductivity of electrolytes, an important contribution to an emerging science. But it's likely that history will remember him most for being the first to propose that burning fossil fuels would eventually hike the planet's temperature. Using early calculations of the rate at which carbon dioxide traps heat in the earth's atmosphere, he concluded that "evaporating our coal mines into the air" would eventually raise the earth's temperature by five or six degrees Celsius, a prediction remarkably close to that now yielded by the most powerful supercomputers on the planet. But Arrhenius, looking at contemporary consumption of coal, thought that doubling the atmospheric concentration of $CO_2$ would take three thousand years; he didn't understand that the century then dawning would see such remarkable economic growth, almost all of it underwritten by fossil fuel.*

# On the Influence of Carbonic Acid in the Air upon the Temperature of the Ground

Svante Arrhenius
*1896*

## Geological Consequences

I should certainly not have undertaken these tedious calculations if an extraordinary interest had not been connected with them. In the Physical Society of Stockholm there have been occasionally very lively discussions on the probable causes of the Ice Age; and these discussions have, in my opinion, led to the conclusion that there exists as yet no satisfactory hypothesis that could explain how the climatic conditions for an ice age could be realized in so short a time as that which has elapsed from the days of the glacial epoch. The common view hitherto has been that the earth has cooled in the lapse of time; and if one did not know that the reverse has been the case, one would certainly assert that this cooling must go on continuously. Conversations with my friend and colleague Professor Högbom, together with the discussions above referred to, led me to make a preliminary estimate of the probable effect of a variation of the atmospheric carbonic acid on the temperature of the earth. As this estimation led to the belief that one might in this way probably find an explanation for temperature variations of five to ten degrees Celsius, I worked out the calculation more in detail, and lay it now before the public and the critics.

From geological researches the fact is well established that in Tertiary times there existed a vegetation and an animal life in the temperate and arctic zones that must have been conditioned by a much higher temperature than the present in the same regions.[1] The temperature in the arctic zones appears to have exceeded the present temperature by about eight or nine degrees. To this genial time the ice age succeeded, and this was one or more times interrupted by interglacial periods with a climate of about the same character as the present, sometimes even milder. When the ice age had its greatest extent, the countries that now enjoy the highest civilization were covered with ice. This was the case with Ireland, Britain (except a small part in the south), Holland, Denmark, Sweden and Norway, Russia (to Kiev, Orel, and Nijni Novgorod), Germany and Austria (to the Harz, Erz-Gebirge, Dresden, and Cracow). At the same time an ice-cap from the Alps covered Switzerland, parts of France, Bavaria south of the Danube, the Tyrol, Styria, and other Austrian countries, and descended into the northern part of Italy. Simultaneously, too, North America was covered with ice on the west coast to the forty-seventh parallel, on the east coast to the fortieth, and in the central part to the thirty-seventh (the confluence of the Mississippi and Ohio rivers). In the most different parts of the world, too, we have found traces of a great ice age, as in the Caucasus, Asia Minor, Syria, the Himalayas, India, Thian Shan, Altai, Atlas, on Mount Kenia and Kilimandjaro (both very near to the equator), in South Africa, Australia, New Zealand, Kerguelen, Falkland Islands, Patagonia and other parts of South America. The geologists in general are inclined to think that these glaciations were simultaneous on the whole earth;[2] and this most natural view would probably have been generally accepted, if the theory of Croll, which demands a genial age on the Southern hemisphere at the same time as an ice age on the Northern and vice versa, had not influenced opinion. By measurements of the displacement of the snow-line we arrive at the result—and this is very concordant for different places—that the temperature at that time must have been four to five degrees Celsius lower than at present. The last glaciation must have taken place in rather recent times, geologically speaking, so that the human race certainly had appeared at that period. Certain American geologists hold the opinion that since the close of the ice age only some 7,000 to 10,000 years have elapsed, but this most probably is greatly underestimated.

One may now ask, How much must the carbonic acid vary according to our figures, in order that the temperature should attain the same values as in the Tertiary and Ice ages respectively? A simple calculation shows that the temperature in the arctic regions would rise about eight to nine degrees Celsius, if the carbonic acid increased to two and a half or three times its present value. In order to get the temperature of the ice age between the fortieth and fiftieth parallels, the carbonic acid in the air should sink to 0.62–0.55 of its present value (lowering of temperature four to five degrees Celsius). The demands of the geologists, that at the genial epochs the climate should be more uniform than now, accords very well with our theory. The geographical annual and diurnal ranges of temperature would be partly smoothed away, if the quantity of carbonic acid was augmented. The reverse would be the case (at least to a latitude of fifty degrees from the equator), if the carbonic acid diminished in amount. But in both these cases I incline to think that the secondary action due to the regress or the progress of the snow-covering would play the most important role. The theory demands also that, roughly speaking, the whole earth should have undergone about the same variations of temperature, so that according to it genial or glacial epochs must have occurred simultaneously on the whole earth. Because of the greater nebulosity of the Southern hemisphere, the variations must there have been a little less (about fifteen percent) than in the Northern hemisphere. The ocean currents, too, must there, as at the present time, have effaced the differences in temperature at different latitudes to a greater extent than in the Northern hemisphere. This effect also results from the greater nebulosity in the arctic zones than in the neighbourhood of the equator.

There is now an important question which should be answered, namely: Is it probable that such great variations in the quantity of carbonic acid as our theory requires have occurred in relatively short geological times? The answer to this question is given by Professor Högbom. As his memoir on this question may not be accessible to most readers of these pages, I have summed up and translated his utterances which are of most importance to our subject:[3]

Although it is not possible to obtain exact quantitative expressions for the reactions in nature by which carbonic acid is developed or consumed, nevertheless there are some factors, of which one may get an approximately true estimate, and from which certain conclusions that throw light on the question may be drawn. In the first place, it seems to be of importance to compare the quantity of carbonic acid now present in the air with the quantities that are being transformed. If the former is insignificant in comparison with the latter, then the probability for variations is wholly other than in the opposite case.

On the supposition that the mean quantity of carbonic acid in the air reaches 0.03 volume percent, this number represents 0.045 percent, by weight, or 0.342 millimeter partial pressure, or 0.466 grams of carbonic acid for every square centimeter of the earth's surface. Reduced to carbon, this quantity would give a layer of about one millimeter thickness over the earth's surface. The quantity of carbon that is fixed in the living organic world can certainly not be estimated with the same degree of exactness; but it is evident that the numbers that might express this quantity ought to be of the same order of magnitude, so that the carbon in the air can neither be conceived of as very great nor as very little, in comparison with the quantity of carbon occurring in organisms. With regard to the great rapidity with which the transformation in organic nature proceeds, the disposable quantity of carbonic acid is not so excessive that changes caused by climatological or other reasons in the velocity and value of that transformation might be not able to cause displacements of the equilibrium.

The following calculation is also very instructive for the appreciation of the relation between the quantity of carbonic acid in the air and the quantities that are transformed. The world's present production of coal reaches in round numbers 500 millions of tons* per annum, or 1 ton per square kilometer of the earth's surface. Transformed into carbonic acid, this quantity would correspond to about a thousandth part of the carbonic acid in the atmosphere. It represents a layer of limestone of 0.003 millimeter thickness over the whole globe, or 1.5 cubic kilometers in cubic measure. This quantity of carbonic acid, which is supplied to the atmosphere chiefly by modern industry, may be regarded as completely compensating the quantity of carbonic acid that is consumed in the formation of limestone (or other mineral carbonates) by the weathering or decomposition of silicates. From

---

*Editor's note: the "ton" referred to here is a metric ton, or 1,000 kg., not an American ton, or 2,000 lbs.

the determination of the amounts of dissolved substances, especially carbonates, in a number of rivers in different countries and climates, and of the quantity of water flowing in these rivers and of their drainage-surface compared with the land-surface of the globe, it is estimated that the quantities of dissolved carbonates that are supplied to the ocean in the course of a year reach at most the bulk of 3 cubic kilometers. As it is also proved that the rivers the drainage regions of which consist of silicates convey very unimportant quantities of carbonates compared with those that flow through limestone regions, it is permissible to draw the conclusion, which is also strengthened by other reasons, that only an insignificant part of these 3 cubic kilometers of carbonates is formed directly by decomposition of silicates. In other words, only an unimportant part of this quantity of carbonate of lime can be derived from the process of weathering in a year. Even though the number given were on account of inexact or uncertain assumptions erroneous to the extent of fifty percent, or more, the comparison instituted is of very great interest, as it proves that the most important of all the processes by means of which carbonic acid has been removed from the atmosphere in all times, namely the chemical weathering of siliceous minerals, is of the same order of magnitude as a process of contrary effect, which is caused by the industrial development of our time, and which must be conceived of as being of a temporary nature.

In comparison with the quantity of carbonic acid which is fixed in limestone (and other carbonates), the carbonic acid of the air vanishes. With regard to the thickness of sedimentary formations and the great part of them that is formed by limestone and other carbonates, it seems not improbable that the total quantity of carbonates would cover the whole earth's surface to a height of hundreds of meters. If we assume 100 meters—a number that may be inexact in a high degree, but probably is underestimated —we find that about 25,000 times as much carbonic acid is fixed to lime in the sedimentary formations as exists free in the air. Every molecule of carbonic acid in this mass of limestone has, however, existed in and passed through the atmosphere in the course of time. Although we neglect all other factors which may have influenced the quantity of carbonic acid in the air, this number lends but very slight probability to the hypothesis, that this quantity should in former geological epochs have changed within limits which do not differ much from the present amount. As the process of weathering has consumed quantities of carbonic acid many thousand times greater than the amount now disposable in the air, and as this process from different geographical, climatological and other causes has in all likelihood proceeded with very different intensity at different epochs, the probability of important variations in the quantity of carbonic acid seems to be very great, even if we take into account the compensating processes which, as we shall see in what follows, are called forth as soon as, for one reason or another, the production or consumption of carbonic acid tends to displace the equilibrium to any considerable degree. One often hears the opinion expressed, that the quantity of carbonic acid in the air ought to have been very much greater formerly than now, and that the diminution should arise from the circumstance that carbonic acid has been taken from the air and stored in the earth's crust in the form of coal and carbonates. In many cases this hypothetical diminution is ascribed only to the formation of coal, whilst the much more important formation of carbonates is wholly overlooked. This whole method of reasoning on a continuous diminution of the carbonic acid in the air loses all foundation in fact, notwithstanding that enormous quantities of carbonic acid in the course of time have been fixed in carbonates, if we consider more closely the processes by means of which carbonic acid has in all times been supplied to the atmosphere. From these we may well conclude that enormous variations have occurred, but not that the variation has always proceeded in the same direction.

Carbonic acid is supplied to the atmosphere by the following processes: (1) volcanic exhalations and geological phenomena connected therewith; (2) combustion of carbonaceous meteorites in the higher regions of the atmosphere; (3) combustion and decay of organic bodies; (4) decomposition of carbonates; (5) liberation of carbonic acid mechanically inclosed in minerals on their fracture or decomposition. The carbonic acid of the air is consumed chiefly by the following processes: (6) formation of carbonates from silicates on weathering; and (7) the consumption of carbonic acid by vegetative processes. The ocean, too, plays an important role as a regulator of the quantity of carbonic acid in the air by means of the absorptive power of its water, which gives off carbonic acid as its temperature rises, and absorbs it as it cools. The processes named under (4) and (5) are of little significance, so that they may be omitted. So too the processes (3) and (7), for the circulation of matter in the organic world goes on so rapidly that their variations cannot have any sensible influence. From this we must except periods in which great quantities of organisms were stored up in sedimentary formations and thus subtracted from the circulation, or in which such stored-up products were, as now, introduced anew into the circulation. The source of carbonic acid named in (2) is wholly incalculable.

Thus the processes (1), (2), and (6) chiefly remain as balancing each other. As the enormous quantities of carbonic acid (representing a pressure of many atmospheres) that are now fixed in the limestone of the earth's crust cannot be conceived to have existed in the air but as an insignificant fraction of the whole at any one time since organic life appeared on the globe, and since therefore the consumption through weathering and formation of carbonates must have been compensated by means of continuous supply, we must regard volcanic exhalations as the chief source of carbonic acid for the atmosphere.

But this source has not flowed regularly and uniformly. Just as single volcanoes have their periods of variation with alternating relative rest and intense activity, in the same manner the globe as a whole seems in certain geological epochs to have exhibited a more violent and general volcanic activity, whilst other epochs have been marked by a comparative quiescence of the volcanic forces. It seems therefore probable that the quantity of carbonic acid in the air has undergone nearly simultaneous variations, or at least that this factor has had an important influence.

If we pass the above-mentioned processes for consuming and producing carbonic acid under review, we find that they evidently do not stand in such a relation to or dependence on one another that any probability exists for the permanence of an equilibrium of the carbonic acid in the atmosphere. An increase or decrease of the supply continued during geological periods must, although it may not be important, conduce to remarkable alterations of the quantity of carbonic acid in the air, and there is no conceivable hindrance to imagining that this might in a certain geological period have been several times greater, or on the other hand considerably less, than now.

As the question of the probability of quantitative variation of the carbonic acid in the atmosphere is in the most decided manner answered by Professor Högbom, there remains only one other point to which I wish to draw attention in a few words, namely: Has no one hitherto proposed any acceptable explanation for the occurrence of genial and glacial periods? Fortunately, during the progress of the foregoing calculations, a memoir was published by the distinguished Italian meteorologist L. De Marchi which relieves me from answering the last question.[4] He examined in detail the different theories hitherto proposed—astronomical, physical, or geographical, and of these I here give a short resume. These theories assert that the occurrence of genial or glacial epochs should depend on one or other change in the following circumstances:

1. The temperature of the earth's place in space.
2. The sun's radiation to the earth (solar constant).
3. The obliquity of the earth's axis to the ecliptic.
4. The position of the poles on the earth's surface.
5. The form of the earth's orbit, especially its eccentricity (Croll).
6. The shape and extension of continents and oceans.
7. The covering of the earth's surface (vegetation).
8. The direction of the oceanic and aërial currents.
9. The position of the equinoxes.

De Marchi arrives at the conclusion that all these hypotheses must be rejected. On the other hand, he is of the opinion that a change in the transparency of the atmosphere would possibly give the desired effect. According to his calculations,

> a lowering of this transparency would effect a lowering of the temperature on the whole earth, slight in the equatorial regions, and increasing with the latitude into the seventieth parallel, nearer the poles again a little less. Further, this lowering would, in non-tropical regions,

be less on the continents than on the ocean and would diminish the annual variations of the temperature. This diminution of the air's transparency ought chiefly to be attributed to a greater quantity of aqueous vapor in the air, which would cause not only a direct cooling but also copious precipitation of water and snow on the continents. The origin of this greater quantity of water-vapor is not easy to explain.

De Marchi has arrived at wholly other results than myself, because he has not sufficiently considered the important quality of selective absorption which is possessed by aqueous vapor. And, further, he has forgotten that if aqueous vapor is supplied to the atmosphere, it will be condensed till the former condition is reached, if no other change has taken place. As we have seen, the mean relative humidity between the fortieth and sixtieth parallels on the northern hemisphere is seventy-six percent. If, then, the mean temperature sank from its actual value +5.3 by four to five degrees Celsius, i.e. to +1.3 or +0.3, and the aqueous vapor remained in the air, the relative humidity would increase to 101 or 105 percent. This is of course impossible, for the relative humidity cannot exceed 100 percent in the free air. *A fortiori* it is impossible to assume that the absolute humidity could have been greater than now in the glacial epoch.

As the hypothesis of Croll still seems to enjoy a certain favor with English geologists, it may not be without interest to cite the utterance of De Marchi on this theory, which he, in accordance with its importance, has examined more in detail than the others. He says, and I entirely agree with him on this point: "Now I think I may conclude that from the point of view of climatology or meteorology, in the present state of these sciences, the hypothesis of Croll seems to be wholly untenable as well in its principles as in its consequences."[5]

It seems that the great advantage which Croll's hypothesis promised to geologists, viz. of giving them a natural chronology, predisposed them in favor of its acceptance. But this circumstance, which at first appeared advantageous, seems with the advance of investigation rather to militate against the theory, because it becomes more and more impossible to reconcile the chronology demanded by Croll's hypothesis with the facts of observation.

I trust that after what has been said the theory proposed in the foregoing pages will prove useful in explaining some points in geological climatology which have hitherto proved most difficult to interpret.

The fear that $CO_2$ might warm the planet remained dormant for most of the first half of the twentieth century while the West was caught up in a pair of World Wars and the rapid spread of industrialization. But changing atmospheric conditions aroused the interest of G. S. Callendar, an English inventor and "steam technologist" for the British Electrical and Allied Industries Research Association, who was also an amateur meteorologist. He collected temperature records from spots around the world (not a simple task even today, but almost unimaginable in 1938) and concluded that the planet's temperature was increasing; he also insisted that $CO_2$ levels in the atmosphere were rising, and he correlated the two effects. Like Arrhenius, he was optimistic about the outcome, thinking that this warmer climate would hold at bay the "deadly glaciers" of another ice age. But his prescience in recognizing what we now call the greenhouse effect was noted by subsequent scientists: into the 1960s, it was often referred to as the Callendar Effect.

# The Artificial Production of Carbon Dioxide and Its Influence on Temperature

G. S. Callendar
*1938*

*Summary*

By fuel combustion man has added about 150,000 million tons of carbon dioxide to the air during the past half century. The author estimates from the best available data that approximately three-quarters of this has remained in the atmosphere.

The radiation absorption coefficients of carbon dioxide and water vapor are used to show the effect of carbon dioxide on "sky radiation." From this the increase in mean temperature, due to the artificial production of carbon dioxide, is estimated to be at the rate of 0.003 degrees Celsius per year at the present time.

The temperature observations at 200 meteorological stations are used to show that world temperatures have actually increased at an average rate of 0.005 degrees Celsius per year during the past half century.

---

Few of those familiar with the natural heat exchanges of the atmosphere, which go into the making of our climates and weather, would be prepared to admit that the activities of man could have any influence upon phenomena of so vast a scale.

In the following paper I hope to show that such influence is not only possible, but is actually occurring at the present time.

It is well known that the gas carbon dioxide has certain strong absorption bands in the infra-red region of the spectrum, and when this fact was discovered some seventy years ago it soon led to speculation on the effect which changes in the amount of the gas in the air could have on the temperature of the earth's surface. In view of the much larger quantities and absorbing power of atmospheric water vapor it was concluded that the effect of carbon dioxide was probably negligible, although certain experts, notably Svante Arrhenius and T. C. Chamberlin, dissented from this view.

Of recent years much new knowledge has been accumulated which has a direct bearing upon this problem, and it is now possible to make a reasonable estimate of the effect of carbon dioxide on temperatures, and also of the rate at which the gas accumulates in the atmosphere. Amongst important factors in such calculations may be mentioned the temperature-pressure-alkalinity-$CO_2$ relation for sea water, determined by C. J. Fox (1909), the vapor pressure–atmospheric radiation relation, observed by A. Angstrom (1918) and others, the absorption spectrum of atmospheric water vapor, observed by Fowle (1918), and a full knowledge of the thermal structure of the atmosphere.

This new knowledge has been used in arriving at the conclusions stated in this paper, but for obvious reasons only those parts having a meteorological character will be referred to here.

*The rate of accumulation of atmospheric carbon dioxide*

I have examined a very accurate set of observations (Brown and Escombe, 1905), taken about the year 1900, on the amount of carbon dioxide in the free air, in relation to the weather maps of the period. From them I concluded that the amount of carbon dioxide in the free air of the North Atlantic region, at the beginning of this century, was $2.74 \pm 0.05$ parts in 10,000 by volume of dry air.

A great many factors which influence the carbon cycle in nature have been examined in order to determine the quantitative relation between the natural movements of this gas and the amounts produced by the combustion of fossil fuel. Such factors included the organic deposit of carbon in swamps, etc., the average rate of fixation of the gas by the carbonization of alkalies from igneous rocks, and so on. The general conclusion from a somewhat lengthy investigation on the natural movements of carbon dioxide was that there is no geological evidence to show that the *net* offtake of the gas is more than a small fraction of the quantity produced from fuel. (The artificial production at present is about 4,500 million tons per year.)

The effect of solution of the gas by the sea water was next considered, because the sea acts as a giant regulator of carbon dioxide and holds some sixty times as much as the atmosphere. The rate at which the sea water could correct an excess of atmospheric carbon dioxide depends mainly upon the fresh volume of water exposed to the air each year, because equilibrium with the atmospheric gases is only established to a depth of about 200 meters during such a period.

The vertical circulation of the oceans is not well understood, but several factors point to an equilibrium time, in which the whole sea volume is exposed to the atmosphere, of between two and five thousand years. Using Fox's solution coefficients for sea water of known total alkalinity and average surface temperature, it is possible to calculate the change in atmospheric $CO_2$ pressure over a given period, when the rate of addition of the gas is known and the equilibrium time for the sea water is assumed. A few such figures are given in Table I, and it will be seen that when periods of a few centuries are considered the sea equilibrium time is less important.

TABLE I. THE EFFECT OF THE ARTIFICIAL PRODUCTION OF CARBON DIOXIDE UPON ITS PRESSURE IN THE ATMOSPHERE.

Annual net addition of $CO_2$ to the air = 4,300 million tons.
Total pressure from $CO_2$ = 0.000274 atmospheres in the year 1900.
Sea surface at 15°C. and total alkalinity = 40 mg. of negative hydroxyl ions per liter; it is this quantity which is maintained neutral by the dissolved $CO_2$. $P(CO_2)$ stands for the pressure of $CO_2$ in the air at normal barometric pressure.

| Sea equilibrium time, years | Date: 1936 | 2000 | 2100 | 2200 |
| --- | --- | --- | --- | --- |
| | | $P(CO_2)$ in atmos./10,000 | | |
| 2000 | 2.89 | 3.14 | 3.46 | 3.73 |
| 5000 | 2.90 | 3.17 | 3.58 | 3.96 |
| All $CO_2$ to the air | 2.96 | 3.35 | 3.96 | 4.58 |

From 1900 to 1936 the increase should be close to six percent.

Since calculating the figures in Table I, I have seen a report of a great number of observations on atmospheric $CO_2$, taken recently in the eastern U.S.A. The mean of 1,156 "free air" readings taken in the years 1930 to 1936 was 3.10 parts in 10,000 by volume. For the measurements at Kew in 1898 to 1901 the mean of ninety-two free air values was 2.92, including a number of rather high values effected by local combustion, etc., and assuming that a similar proportion of the American readings are affected in the same way, the difference is equal to an increase of six percent. Such close agreement with the calculated increase is, of course, partly accidental.

[. . .]

As regards the long period temperature variations represented by the Ice Ages of the geologically recent past, I have made many calculations to see if the natural movements of carbon dioxide could be rapid enough to account for the great changes of the amount in the atmosphere which would be necessary to give glacial periods with a duration of about 30,000 years. I find it almost impossible to account for movements of the gas of the required order because of the almost inexhaustible supply from the oceans, when its pressure in the air becomes low enough to give a fall of five to eight degrees Celsius in mean temperatures. Of course, if the effect of carbon dioxide on temperatures was considerably greater than supposed, glacial periods might well be accounted for in this way.

In conclusion it may be said that the combustion of fossil fuel, whether it be peat from the surface or oil from 10,000 feet below, is likely to prove beneficial to mankind in several ways, besides the provision of heat and power. For instance the above mentioned small increases of mean temperature would be important at the northern margin of cultivation, and the growth of favourably situated plants is directly proportional to the carbon dioxide pressure (Brown and Escombe, 1905). In any case the return of the deadly glaciers should be delayed indefinitely.

As regards the reserves of fuel these would be sufficient to give at least ten times as much carbon dioxide as there is in the air at present.

*If there's a single paper that could be considered the opening volley of the climate wars, this is it. Until this moment, few scientists had taken seriously the work of Arrhenius and Callendar because of a widespread belief that the planet's vast oceans could absorb however much $CO_2$ humans produced. Roger Revelle and Hans Suess challenged that assumption. Revelle was one of the greatest oceanographers of his age, and a powerful force in postwar science; Suess was a chemist fascinated by the newly available tool of carbon dating, a technique that allows scientists to estimate the age of ancient organic material (including carbon in the oceans). Both worked out of the Scripps Institution of Oceanography in La Jolla, California. What their 1956 paper shows, essentially, is that seawater is already saturated with $CO_2$, and consequently that the ocean's particular chemistry would keep it from being an eternal sink for the vast quantities of the gas that humans were now producing. They speculated that $CO_2$ must therefore be accumulating in the atmosphere, and hence that "human beings are now carrying out a large scale geophysical experiment of a kind that could not have happened in the past nor be reproduced in the future." That's one of the most quoted sentences in the history of climate change. Given what we now know, it carries a dire tone. But in the late 1950s "experiment" didn't sound quite so sinister; Revelle and Suess were, above all, scientists.*

# Carbon Dioxide Exchange between Atmosphere and Ocean and the Question of an Increase of Atmospheric $CO_2$ during the Past Decades

Roger Revelle and Hans E. Suess
1957

## Abstract

From a comparison of $C^{14}/C^{12}$ and $C^{13}/C^{12}$ ratios in wood and in marine material and from a slight decrease of the $C^{14}$ concentration in terrestrial plants over the past fifty years it can be concluded that the average lifetime of a $CO_2$ molecule in the atmosphere before it is dissolved into the sea is of the order of ten years. This means that most of the $CO_2$ released by artificial fuel combustion since the beginning of the industrial revolution must have been absorbed by the oceans. The increase of atmospheric $CO_2$ from this cause is at present small but may become significant during future decades if industrial fuel combustion continues to rise exponentially.

Present data on the total amount of $CO_2$ in the atmosphere, on the rates and mechanisms of exchange, and on possible fluctuations in terrestrial and marine organic carbon, are inadequate for accurate measurement of future changes in atmospheric $CO_2$.

Contribution from the Scripps Institution of Oceanography, New Series, No. 900. This paper represents in part results of research carried out by the University of California under contract with the Office of Naval Research.

An opportunity exists during the International Geophysical Year to obtain much of the necessary information.

## Introduction

In the middle of the nineteenth century appreciable amounts of carbon dioxide began to be added to the atmosphere through the combustion of fossil fuels. The rate of combustion has continually increased so that at the present time the annual increment from this source is nearly 0.4 percent of the total atmospheric carbon dioxide. By 1960 the amount added during the past century will be more than fifteen percent.

Callendar (1938, 1940, 1949) believed that nearly all the carbon dioxide produced by fossil fuel combustion has remained in the atmosphere, and he suggested that the increase in atmospheric carbon dioxide may account for the observed slight rise of average temperature in northern latitudes during recent decades. He thus revived the hypothesis of T.C. Chamberlin (1899) and S. Arrhenius (1903) that climatic changes may be related to fluctuations in the carbon dioxide content of the air. These authors supposed that an increase of carbon dioxide in the upper atmosphere would lower the mean level of back radiation in the infrared and thereby increase the average temperature near the earth's surface.

Subsequently, other authors have questioned Callendar's conclusions on two grounds. First, comparison of measurements made in the nineteenth century and in recent years do not demonstrate that there has been a significant increase in atmospheric $CO_2$ (Slocum, 1955; Fonselius *et. al.*, 1956). Most of the excess $CO_2$ from fuel combustion may have been transferred to the ocean, a possibility suggested by S. Arrhenius (1903). Second, a few percent increase in the $CO_2$ content of the air, even if it has occurred, might not produce an observable increase in average air temperature near the ground in the face of fluctuations due to other causes. So little is known about the thermodynamics of the atmosphere that it is not certain whether or how a change in infrared back radiation from the upper air would affect the temperature near the surface. Calculations by Plass (1956) indicate that a ten percent increase in atmospheric carbon dioxide would increase the average temperature by 0.36 degrees Celsius. But, amplifying or feed-back processes may exist such that a slight change in the character of the back radiation might have a more pronounced effect. Possible examples are a decrease in albedo of the earth due to melting of ice caps or a rise in water vapor content of the atmosphere (with accompanying increased infrared absorption near the surface) due to increased evaporation with rising temperature.

During the next few decades the rate of combustion of fossil fuels will continue to increase, if the fuel and power requirements of our world-wide industrial civilization continue to rise exponentially, and if these needs are met only to a limited degree by development of atomic power. Estimates by the UN [United Nations] (1955) indicate that during the first decade of the twenty-first century fossil-fuel combustion could produce an amount of carbon dioxide equal to twenty percent of that now in the atmosphere [. . .]. This is probably two orders of magnitude greater than the usual rate of carbon dioxide production from volcanoes, which on the average must be equal to the rate at which silicates are weathered to carbonates [. . .]. Thus human beings are now carrying out a large scale geophysical experiment of a kind that could not have happened in the past nor be reproduced in the future. Within a few centuries we are returning to the atmosphere and oceans the concentrated organic carbon stored in sedimentary rocks over hundreds of millions of years. This experiment, if adequately documented, may yield a far-reaching insight into the processes determining weather and climate. It therefore becomes of prime importance to attempt to determine the way in which carbon dioxide is partitioned between the atmosphere, the oceans, the biosphere and the lithosphere.

This may be the most famous image of the climate change era. After Revelle and Suess calculated that $CO_2$ must be accumulating in the atmosphere because it wasn't building up in the oceans, the next logical question was, how fast? The approaching International Geophysical Year (IGY) in 1957 gave them a chance to find out. They directed money from the IGY budget toward a careful young scientist named Dave Keeling, who bought and modified infrared gas analyzers and placed them at several locations around the planet, most importantly the Hawaiian volcano Mauna Loa. There he observed for the first time the yearly fluctuation in $CO_2$ levels that occurred as Northern Hemisphere vegetation absorbed $CO_2$ while growing during the spring and then released it while dying in the fall—it was as if he could see the planet breathing. Even more importantly, within a very few years he could see that the annual maximum value for $CO_2$ was steadily rising, a pattern that has now continued unabated for five decades. When Keeling turned on his machine in March of 1958, the earth's atmosphere was 313 parts per million $CO_2$; today, that number has passed 392. Upon Keeling's death in 2005, one colleague said there had been three historical instances in which "dedication to scientific measurements has changed all of science": Tycho Brahe's careful observation of planetary orbits, which laid the ground for Newton's discoveries; Albert Michelson's measurement of the speed of light, which allowed Einstein's work on relativity; and Keeling's famous curve, "the single most important environmental data set taken in the twentieth century."

# The Keeling Curve

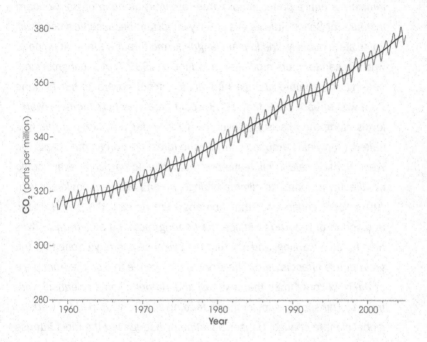

Courtesy the NASA Earth Observatory. NASA graph by Robert Simmon, based on data provided by the NOAA Climate Monitoring and Diagnostics Laboratory.

*If the scientific story of global warming has one great hero, he is James Hansen, and not only because he is the most important climatologist of his era, whose massive computer models were demonstrating by the early 1980s that increased $CO_2$ posed a dire threat. Hansen—who has spent (despite attempts by several administrations to fire or muzzle him) almost his entire professional career at NASA as head of the Goddard Institute for Space Studies—is equally important because he's been willing to state publicly, and in plain language, the threat we face. At no point was this candor more crucial than in June of 1988. Earlier congressional hearings on climate change had yielded scant coverage, but by June 23 it was already clear that 1988 would be a very hot summer—water levels along the Mississippi eventually fell so far that barge traffic was halted; crop yields dropped sharply; and forest fires broke out across the West. Hansen said in his testimony that while no particular event could be directly attributed to climate change, he and his team could say with "99 percent confidence" that humans were heating the planet. Later, in a scrum of reporters outside the hearing room, he added that it was time to "stop waffling, and say that the evidence is pretty strong that the greenhouse effect is here." The story ran above the fold on the front page of the* New York Times *the next day, and although some scientists were irked by Hansen's outspokenness, funding suddenly jumped for work on global climate change. Hansen's testimony had ignited the most intense period of scientific investigation of any topic ever.*

# Statement of Dr. James Hansen

James Hansen
*1988*

*Dr. Hansen:* Mr. Chairman and committee members, thank you for the opportunity to present the results of my research on the greenhouse effect which has been carried out with my colleagues at the NASA Goddard Institute for Space Studies.

I would like to draw three main conclusions. Number one, the earth is warmer in 1988 than at any time in the history of instrumental measurements. Number two, the global warming is now large enough that we can ascribe with a high degree of confidence a cause and effect relationship to the greenhouse effect. And number three, our computer climate simulations indicate that the greenhouse effect is already large enough to begin to effect the probability of extreme events such as summer heat waves.

My first viewgraph, which I would like to ask Suki to put up if he would, shows the global temperature over the period of instrumental records which is about 100 years. The present temperature is the highest in the period of record. The rate of warming in the past twenty-five years, as you can see on the right, is the highest on record. The four warmest years, as the Senator mentioned, have all been in the 1980s. And 1988 so far

Fig. 1

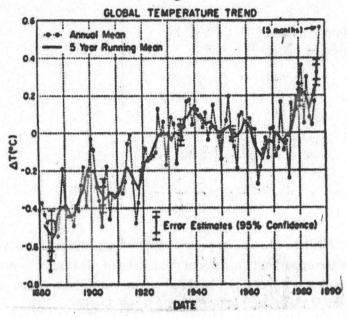

Fig. 2

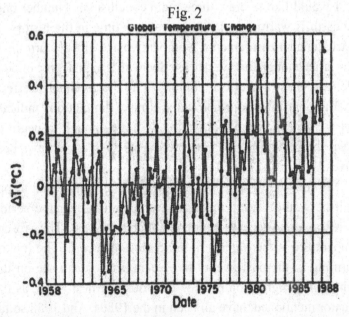

is so much warmer than 1987, that barring a remarkable and improbable cooling, 1988 will be the warmest year on the record.

Now let me turn to my second point which is causal association of the greenhouse effect and the global warming. Causal association requires first that the warming be larger than natural climate variability and, second, that the magnitude and nature of the warming be consistent with the greenhouse mechanism. These points are both addressed on my second viewgraph. The observed warming during the past thirty years, which is the period when we have accurate measurements of atmospheric composition, is shown by the heavy black line in this graph. The warming is almost 0.4 degrees Centigrade by 1987 relative to climatology, which is defined as the thirty-year mean, 1950 to 1980 and, in fact, the warming is more than 0.4 degrees Centigrade in 1988. The probability of a chance warming of that magnitude is about one percent. So, with ninety-nine percent confidence we can state that the warming during this time period is a real warming trend.

The other curves in this figure are the results of global climate model calculations for three scenarios of atmospheric trace gas growth. We have considered several scenarios because there are uncertainties in the exact trace gas growth in the past and especially in the future. We have considered cases ranging from

---

Fig. 1. Global surface air temperature change for the past century, with the zero point defined as the 1951–1980 mean. Uncertainty bars (95% confidence limits) are based on an error analysis as described in reference 6; inner bars refer to the five-year mean and outer bars to the annual mean. The analyzed uncertainty is a result of incomplete spatial coverage by measurement stations, primarily in ocean areas. The 1988 point compares the January–May 1988 temperature to the mean for the same five months in 1951–1980.

Fig. 2. Global surface air temperature change at seasonal resolution for the past thirty years. Figure 1 and 2 are updates of results in reference 6.

business as usual, which is Scenario A, to draconian emission cuts, Scenario C, which would totally eliminate net trace gas growth by year 2000.

The main point to be made here is that the expected global warming is of the same magnitude as the observed warming. Since there is only a one-percent chance of an accidental warming of this magnitude, the agreement with the expected greenhouse effect is of considerable significance. Moreover, if you look at the next level of detail in the global temperature change, there are clear signs of the greenhouse effect. Observational data suggests a cooling in the stratosphere while the ground is warming. The data suggest somewhat more warming over land and sea ice regions than over open ocean, more warming at high latitudes than at low latitudes, and more warming in the winter than in the summer. In all of these cases, the signal is at best just beginning to emerge, and we need more data. Some of these details, such as the northern hemisphere high-latitude temperature trends, do not look exactly like the greenhouse effect, but that is expected. There are certainly other climate change factors involved in addition to the greenhouse effect.

Altogether the evidence that the earth is warming by an amount which is too large to be a chance fluctuation and the similarity of the warming to that expected from the greenhouse effect represents a very strong case. In my opinion, that the greenhouse effect has been detected, and it is changing our climate now.

Then my third point. Finally, I would like to address the question of whether the greenhouse effect is already large enough to affect the probability of extreme events, such as summer heat waves. As shown in my next viewgraph, we have used the temperature changes computed in our global climate model to estimate the impact of the greenhouse effect on the frequency of hot summers in Washington, D.C. and Omaha, Nebraska. A hot summer is defined as the hottest one-third of the summers in the 1950 to 1980 period, which is the period the Weather Bureau uses for defining climatology. So, in that period the probability of having a hot summer was thirty-three percent, but by the 1990s, you can see that the greenhouse effect has increased the probability of a hot summer to somewhere between fifty-five and seventy percent in Washington according to our climate model simulations. In the late 1980s, the probability of a hot summer would be somewhat less than that. You can interpolate to a value of something like forty to sixty percent.

I believe that this change in the frequency of hot summers is large enough to be noticeable to the average person. So, we have already reached a point that the greenhouse effect is important. It may also have important implications other than for creature comfort.

My last viewgraph shows global maps of temperature anomalies for a particular month, July, for several different years between 1986 and 2029, as computed with our global climate model for the intermediate trace gas Scenario B. As shown by the graphs on the left where yellow and red colors represent areas that are warmer than climatology and blue areas represent areas that are colder than climatology, at the present time in the 1980s the greenhouse warming is smaller than the natural variability of the local temperature. So, in any given month, there is almost as much area that is cooler than normal as there is area warmer than normal. A few decades in the future, as shown on the right, it is warm almost everywhere.

However, the point that I would like to make is that in the late 1980s and in the 1990s we notice a clear tendency in our model for greater than average warming in the southeast United States and the midwest. In our model this result seems to arise because the Atlantic Ocean off the coast of the United States warms more slowly than the land. This leads to high pressure along the east coast and circulation of warm air north into the midwest or the southeast. There is only a tendency for this phenomenon. It is

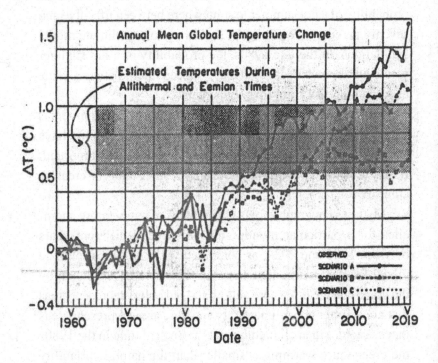

Fig. 3. Annual mean global surface air temperature computed for trace gas scenarios A, B and C described in reference 1. [Scenario A assumes continued growth rates of trace gas emissions typical of the the past twenty years, i.e., about 1.50 yr⁻¹ emission growth; scenario B has emission rates approximately fixed at current rates; scenario C drastically reduces trace gas emissions between 1990 and 2000.] Observed temperatures are from reference 6. The shaded range is an estimate of global temperature during the peak of the current and previous interglacial periods, about 6,000 and 120,000 years before present, respectively. The zero point for observations is the 1951–1980 mean (reference 6); the zero point for the model is the control run mean.

certainly not going to happen every year, and climate models are certainly an imperfect tool at this time. However, we conclude that there is evidence that the greenhouse effect increases the likelihood of heat-wave drought situations in the southeast and midwest United States even though we cannot blame a specific drought on the greenhouse effect.

Therefore, I believe that it is not a good idea to use the period 1950 to 1980 for which climatology is normally defined as an indication of how frequently droughts will occur in the future. If our model is approximately correct, such situations may be more common in the next ten to fifteen years than they were in the period 1950 to 1980.

Finally, I would like to stress that there is a need for improving these global climate models, and there is a need for global observations if we're going to obtain a full understanding of these phenomena.

That concludes my statement, and I'd be glad to answer questions if you'd like.

*In the wake of Hansen's testimony, and the rush toward scientific inves-
tigation that ensued, attention slowly turned to a new group: the Intergovernmental Panel on Climate Change (IPCC). Established in 1988 by the
United Nations, the IPCC involves most of the world's climatologists and
a wide variety of other researchers who study related topics. They regularly issue reports, and in 1995 released the second, and perhaps most
important, of these. After synthesizing the relevant peer-reviewed studies,
the report offered a set of conservative but powerful conclusions. Most
importantly, it insisted that "the balance of evidence suggests that there is
a discernible human influence on global climate." In other words, by 1995
the world's preeminent climate scientists were saying that Hansen was
right: global warming is upon us. That conclusion has grown ever stronger
in subsequent reports, placing the burden of proof increasingly on anyone
attempting to undermine the global warming consensus.*

# Summary for Policymakers: The Science of Climate Change

IPCC Working Group I
*1995*

Considerable progress has been made in the understanding of climate change* science since 1990, and new data and analyses have become available.

## 1. Greenhouse Gas Concentrations Have Continued to Increase

Increases in greenhouse gas concentrations since pre-industrial times (i.e., since about 1750) have led to a positive radiative forcing† of climate, tending to warm the surface and to produce other changes of climate.

• The atmospheric concentrations of greenhouse gases, *inter alia*, carbon dioxide ($CO_2$), methane ($CH_4$) and nitrous oxide

---

* Climate change in IPCC Working Group I usage refers to any change in climate over time whether due to natural variability or as a result of human activity. This differs from the usage in the UN Framework Convention on Climate Change where "climate change" refers to a change of climate which is attributed directly or indirectly to human activity that alters the composition of the global atmosphere and which is in addition to natural climate variability observed over comparable time periods.

† A simple measure of the importance of a potential climate change mechanism. Radiative forcing is the perturbation to the energy balance of the Earth-atmosphere system (in Watts per square metre [$Wm^{-2}$]).

($N_2O$) have grown significantly: by about thirty percent, 145 percent, and fifteen percent, respectively (values for 1992). These trends can be attributed largely to human activities, mostly fossil-fuel use, land-use change and agriculture.

• The growth rates of $CO_2$, $CH_4$ and $N_2O$ concentrations were low during the early 1990s. While this apparently natural variation is not yet fully explained, recent data indicate that the growth rates are currently comparable to those averaged over the 1980s.

• The direct radiative forcing of the long-lived greenhouse gases (2.45 watts per square meter [$Wm^{-2}$]) is due primarily to increases in the concentrations of $CO_2$ (1.56 $Wm^{-2}$), $CH_4$ (0.47 $Wm^{-2}$) and $N_2O$ (0.14 $Wm^{-2}$) (values for 1992).

• Many greenhouse gases remain in the atmosphere for a long time (for $CO_2$ and $N_2O$, many decades to centuries), hence they affect radiative forcing on long time-scales.

• The direct radiative forcing due to the CFCs [chlorofluorocarbons] and HCFCs [hydrochlorofluorocarbons] combined is 0.25 $Wm^{-2}$. However, their net radiative forcing is reduced by about 0.1 $Wm^{-2}$ because they have caused stratospheric ozone depletion which gives rise to a negative radiative forcing.

• Growth in the concentration of CFCs, but not HCFCs, has slowed to about zero. The concentrations of both CFCs and HCFCs, and their consequent ozone depletion, are expected to decrease substantially by 2050 through implementation of the Montreal Protocol and its Adjustments and Amendments.

• At present, some long-lived greenhouse gases (particularly HFCs, a CFC substitute, PFCs and $SF_6$) contribute little to radiative forcing, but their projected growth could contribute several percent to radiative forcing during the twenty-first century.

- If carbon dioxide emissions were maintained at near current (1994) levels, they would lead to a nearly constant rate of increase in atmospheric concentrations for at least two centuries, reaching about 500 parts per million by volume (approaching twice the pre-industrial concentration of 280 ppmv) by the end of the twenty-first century.

- A range of carbon-cycle models indicates that stabilization of atmospheric $CO_2$ concentrations at 450, 650 or 1,000 ppmv could be achieved only if global anthropogenic $CO_2$ emissions drop to 1990 levels by, respectively, approximately forty, 140 or 240 years from now, and drop substantially below 1990 levels subsequently.

- Any eventual stabilized concentration is governed more by the accumulated anthropogenic $CO_2$ emissions from now until the time of stabilization than by the way those emissions change over the period. This means that, for a given stabilized concentration value, higher emissions in early decades require lower emissions later on. Among the range of stabilization cases studied, for stabilization at 450, 650 or 1,000 ppmv, accumulated anthropogenic emissions over the period 1991 to 2100 are 630 gigatonnes of carbon [1 billion tonnes of carbon, or GtC], 1,030 GtC and 1,410 GtC, respectively (plus or minus approximately fifteen percent in each case). For comparison, the corresponding accumulated emissions for IPCC IS92 emission scenarios range from 770 to 2,190 GtC.

- Stabilization of $CH_4$ and $N_2O$ concentrations at today's levels would involve reductions in anthropogenic emissions of eight percent and more than fifty percent respectively.

- There is evidence that tropospheric ozone concentrations in the Northern Hemisphere have increased since pre-industrial times because of human activity and that this has resulted in a positive radiative forcing. This forcing is not yet well characterized, but it is estimated to be about 0.4 watts per square meter (fifteen percent of that from the long-lived greenhouse gases). However, the observations of the most recent decade show that the upward trend has slowed significantly or stopped.

## 2. Anthropogenic Aerosols Tend to Produce Negative Radiative Forcings

- Tropospheric aerosols (microscopic airborne particles) resulting from combustion of fossil fuels, biomass burning and other sources have led to a negative direct forcing of about 0.5 watts per square meter, as a global average, and possibly also to a negative indirect forcing of a similar magnitude. While the negative forcing is focused in particular regions and subcontinental areas, it can have continental to hemispheric scale effects on climate patterns.
- Locally, the aerosol forcing can be large enough to more than offset the positive forcing due to greenhouse gases.
- In contrast to the long-lived greenhouse gases, anthropogenic aerosols are very short-lived in the atmosphere, hence their radiative forcing adjusts rapidly to increases or decreases in emissions.

## 3. Climate Has Changed Over the Past Century

- At any one location, year-to-year variations in weather can be large, but analyses of meteorological and other data over large areas and over periods of decades or more have provided evidence for some important systematic changes.

- Global mean surface air temperature has increased by between about 0.3 and 0.6 degrees Celsius since the late nineteenth century; the additional data available since 1990 and the re-analyses since then have not significantly changed this range of estimated increase.
- Recent years have been among the warmest since 1860, i.e., in the period of instrumental record, despite the cooling effect of the 1991 Mount Pinatubo volcanic eruption.
- Night-time temperatures over land have generally increased more than daytime temperatures.
- Regional changes are also evident. For example, the recent warming has been greatest over the mid-latitude continents in winter and spring, with a few areas of cooling, such as the North Atlantic ocean. Precipitation has increased over land in high latitudes of the Northern Hemisphere, especially during the cold season.
- Global sea level has risen by between ten and twenty-five centimeters over the past 100 years and much of the rise may be related to the increase in global mean temperature.
- There are inadequate data to determine whether consistent global changes in climate variability or weather extremes have occurred over the twentieth century. On regional scales there is clear evidence of changes in some extremes and climate variability indicators (e.g., fewer frosts in several widespread areas; an increase in the proportion of rainfall from extreme events over the contiguous states of the USA). Some of these changes have been toward greater variability; some have been toward lower variability.
- The 1990 to mid-1995 persistent warm-phase of the *El Niñ*-Southern Oscillation (which causes droughts and floods in many areas) was unusual in the context of the last 120 years.

## 4. The Balance of Evidence Suggests a Discernible Human Influence on Global Climate

Any human-induced effect on climate will be superimposed on the background "noise" of natural climate variability, which results both from internal fluctuations and from external causes such as solar variability or volcanic eruptions. Detection and attribution studies attempt to distinguish between anthropogenic and natural influences. "Detection of change" is the process of demonstrating that an observed change in climate is highly unusual in a statistical sense, but does not provide a reason for the change. "Attribution" is the process of establishing cause and effect relations, including the testing of competing hypotheses.

Since the 1990 IPCC Report, considerable progress has been made in attempts to distinguish between natural and anthropogenic influences on climate. This progress has been achieved by including effects of sulphate aerosols in addition to greenhouse gases, thus leading to more realistic estimates of human-induced radiative forcing. These have then been used in climate models to provide more complete simulations of the human-induced climate-change "signal." In addition, new simulations with coupled atmosphere-ocean models have provided important information about decade-to-century time-scale natural internal climate variability. A further major area of progress is the shift of focus from studies of global-mean changes to comparisons of modeled and observed spatial and temporal patterns of climate change.

The most important results related to the issues of detection and attribution are:

- The limited available evidence from proxy climate indicators suggests that the twentieth-century global mean temperature is at least as warm as any other century since at least 1400 A.D. Data prior to 1400 are too sparse to allow the reliable estimation of global mean temperature.

• Assessments of the statistical significance of the observed global mean surface air temperature trend over the last century have used a variety of new estimates of natural internal and externally forced variability. These are derived from instrumental data, palaeodata, simple and complex climate models, and statistical models fitted to observations. Most of these studies have detected a significant change and show that the observed warming trend is unlikely to be entirely natural in origin.

• More convincing recent evidence for the attribution of a human effect on climate is emerging from pattern-based studies, in which the modeled climate response to combined forcing by greenhouse gases and anthropogenic sulphate aerosols is compared with observed geographical, seasonal and vertical patterns of atmospheric temperature change. These studies show that such pattern correspondences increase with time, as one would expect, as an anthropogenic signal increases in strength. Furthermore, the probability is very low that these correspondences could occur by chance as a result of natural internal variability only. The vertical patterns of change are also inconsistent with those expected for solar and volcanic forcing.

• Our ability to quantify the human influence on global climate is currently limited because the expected signal is still emerging from the noise of natural variability, and because there are uncertainties in key factors. These include the magnitude and patterns of long-term natural variability and the time-evolving pattern of forcing by, and response to, changes in concentrations of greenhouse gases and aerosols, and land surface changes. Nevertheless, the balance of evidence suggests that there is a discernible human influence on global climate.

## 5. Climate Is Expected to Continue to Change in the Future

The IPCC has developed a range of scenarios, IS92a-f, of future greenhouse-gas and aerosol-precursor emissions based on assumptions concerning population and economic growth, land-use, technological changes, energy availability, and fuel mix during the period 1990 to 2100. Through understanding of the global carbon cycle and of atmospheric chemistry, these emissions can be used to project atmospheric concentrations of greenhouse gases and aerosols and the perturbation of natural radiative forcing. Climate models can then be used to develop projections of future climate.

- The increasing realism of simulations of current and past climate by coupled atmosphere-ocean climate models has increased our confidence in their use for projection of future climate change. Important uncertainties remain, but these have been taken into account in the full range of projections of global mean temperature and sea-level change.
- For the mid-range IPCC emission scenario, IS92a, assuming the "best estimate" value of climate sensitivity* and including the effects of future increases in aerosol, models project an increase in global mean surface air temperature relative

---

* In IPCC reports, climate sensitivity usually refers to the long-term (equilibrium) change in global mean surface temperature following a doubling of atmospheric equivalent $CO_2$ concentration. More generally, it refers to the equilibrium change in surface air temperature following a unit change in radiative forcing ($^{\circ}C/Wm^{-2}$).

to 1990 of about two degrees Celsius by 2100. This estimate is approximately one-third lower than the "best estimate" in 1990. This is due primarily to lower emission scenarios (particularly for $CO_2$ and the CFCs), the inclusion of the cooling effect of sulphate aerosols, and improvements in the treatment of the carbon cycle. Combining the lowest IPCC emission scenario (IS92c) with a "low" value of climate sensitivity and including the effects of future changes in aerosol concentrations leads to a projected increase of about one degree Celsius by 2100. The corresponding projection for the highest IPCC scenario (IS92e) combined with a "high" value of climate sensitivity gives a warming of about 3.5 degrees Celsius. In all cases the average rate of warming would probably be greater than any seen in the last 10,000 years, but the actual annual-to-decadal changes would include considerable natural variability. Regional temperature changes could differ substantially from the global mean value. Because of the thermal inertia of the oceans, only fifty to ninety percent of the eventual equilibrium temperature change would have been realized by 2100 and temperature would continue to increase beyond 2100, even if concentrations of greenhouse gases were stabilized by that time.

• Average sea level is expected to rise as a result of thermal expansion of the oceans and melting of glaciers and ice-sheets. For the IS92a scenario, assuming the "best estimate" values of climate sensitivity and of ice-melt sensitivity to warming, and including the effects of future changes in aerosol, models project an increase in sea level of about fifty centimeters from the present to 2100. This estimate is approximately twenty-five percent lower than the "best estimate" in 1990 due to the lower temperature projection, but also reflecting improvements in the climate and ice-melt models. Combining the lowest emission scenario (IS92c) with the "low" climate and ice-melt sensitivities and including aerosol effects gives a projected sea-level rise of about 15 centimeters from the present to 2100. The corresponding projection for the highest emission scenario (IS92e) combined with "high" climate and ice-melt sensitivities gives a sea-level rise of about 95 centimeters from the present to 2100. Sea level would continue to rise at a similar rate in future centuries beyond 2100, even if concentrations of greenhouse gases were stabilized by that time, and would continue to do so even beyond the time of stabilization of global mean temperature. Regional sea-level changes may differ from the global mean value owing to land movement and ocean current changes.

- Confidence is higher in the hemispheric-to-continental scale projections of coupled atmosphere-ocean climate models than in the regional projections, where confidence remains low. There is more confidence in temperature projections than hydrological changes.

- All model simulations, whether they were forced with increased concentrations of greenhouse gases and aerosols or with increased concentrations of greenhouse gases alone, show the following features: greater surface warming of the land than of the sea in winter; a maximum surface warming in high northern latitudes in winter, little surface warming over the Arctic in summer; an enhanced global mean hydrological cycle, and increased precipitation and soil moisture in high latitudes in winter. All these changes are associated with identifiable physical mechanisms.

- In addition, most simulations show a reduction in the strength of the north Atlantic thermohaline circulation and a widespread reduction in diurnal range of temperature. These features too can be explained in terms of identifiable physical mechanisms.

- The direct and indirect effects of anthropogenic aerosols have an important effect on the projections. Generally, the magnitudes of the temperature and precipitation changes are smaller when aerosol effects are represented, especially in northern mid-latitudes. Note that the cooling effect of aerosols is not a simple offset to the warming effect of greenhouse gases, but significantly affects some of the continental scale patterns of climate change, most noticeably in the summer hemisphere. For example, models that consider only the effects of greenhouse gases generally project an increase in precipitation and soil moisture in the Asian summer monsoon region, whereas models that include, in addition, some of the effects of aerosols suggest that monsoon precipitation may decrease. The spatial and temporal distribution of aerosols greatly influences regional projections, which are therefore more uncertain.

- A general warming is expected to lead to an increase in the occurrence of extremely hot days and a decrease in the occurrence of extremely cold days.

- Warmer temperatures will lead to a more vigorous hydrological cycle; this translates into prospects for more severe droughts and/or floods in some places and less severe droughts and/or floods in other places. Several models indicate an increase in precipitation intensity, suggesting a possibility for more extreme rainfall events. Knowledge is currently insufficient to say whether there will be any changes in the occurrence or geographical distribution of severe storms, e.g., tropical cyclones.

• Sustained rapid climate change could shift the competitive balance among species and even lead to forest dieback, altering the terrestrial uptake and release of carbon. The magnitude is uncertain, but could be between zero and 200 gigatonnes of carbon over the next one to two centuries, depending on the rate of climate change.

## 6. *There Are Still Many Uncertainties*

Many factors currently limit our ability to project and detect future climate change. In particular, to reduce uncertainties further work is needed on the following priority topics:

• Estimation of future emissions and biogeochemical cycling (including sources and sinks) of greenhouse gasses, aerosols and aerosol precursors and projections of future concentrations and radiative properties.

• Representation of climate processes in models, especially feedbacks associated with clouds, oceans, sea ice and vegetation, in order to improve projections of rates and regional patterns of climate change.

• Systematic collection of long-term instrumental and proxy observations of climate system variables (e.g., solar output, atmospheric energy balance components, hydrological cycles, ocean characteristics and ecosystem changes) for the purposes of model testing, assessment of temporal and regional variability, and for detection and attribution studies.

Future unexpected, large and rapid climate system changes (as have occurred in the past) are, by their nature, difficult to predict. This implies that future climate changes may also involve "surprises." In particular, these arise from the non-linear nature of the climate system. When rapidly forced, non-linear systems are especially subject to unexpected behavior. Progress can be made by investigating non-linear processes and sub-components of the climatic system. Examples of such non-linear behavior include rapid circulation changes in the North Atlantic and feedbacks associated with terrestrial ecosystem changes.

In 2000, the Dutch Nobel Laureate chemist Paul Crutzen and his colleague Eugene F. Stoermer penned an article for the newsletter of the International Geosphere-Biosphere Programme in which they suggested that the Holocene—the geological epoch that marked the 10,000 years of human civilization—had come to an end, to be replaced by the Anthropocene, a world deeply influenced by human activity. This was not a new formulation, at least for nonscientists (see, for instance, the excerpt from 1989's The End of Nature in the Impact section of this anthology), but for scientists such a formulation helped describe the magnitude of the changes they were seeing. Interestingly, Crutzen's Nobel Prize was for his work on the other great environmental problem of the age: the man-made hole in the ozone layer.

# The "Anthropocene"

Paul J. Crutzen and Eugene F. Stoermer
2000

The name Holocene ("Recent Whole") for the post-glacial geological epoch of the past ten to twelve thousand years seems to have been proposed for the first time by Sir Charles Lyell in 1833, and adopted by the International Geological Congress in Bologna in 1885.[1] During the Holocene mankind's activities gradually grew into a significant geological, morphological force, as recognized early on by a number of scientists. Thus, G.P. Marsh already in 1864 published a book with the title "Man and Nature," more recently reprinted as "The Earth as Modified by Human Action".[2] Stoppani in 1873 rated mankind's activities as a "new telluric force which in power and universality may be compared to the greater forces of earth" [quoted from Clark[3]]. Stoppani already spoke of the anthropozoic era. Mankind has now inhabited or visited almost all places on Earth; he has even set foot on the moon.

The great Russian geologist V.I. Vernadsky in 1926 recognized the increasing power of mankind as part of the biosphere with the following excerpt[4] "... the direction in which the processes of evolution must proceed, namely towards increasing consciousness and thought, and forms having greater and greater influence on their surroundings." He, the French Jesuit P. Teilhard de Chardin and E. Le Roy in 1924 coined the term "noösphere," the world of thought, to mark the growing role played by mankind's brainpower and technological talents in shaping its own future and environment.

The expansion of mankind, both in numbers and per-capita exploitation of Earth's resources has been astounding.[5] To give a few examples: During the past three centuries human population increased tenfold to 6 billion, accompanied, e.g., by a growth in cattle population to 1.4 billion[6] (about one cow per average-size family). Urbanization has even increased tenfold in the past century. In a few generations mankind is exhausting the fossil fuels that were generated over several hundred million years. The release of $SO_2$, globally about 160 teragrams [160 million metric tons] per year to the atmosphere by coal and oil burning, is at least two times larger than the sum of all natural emissions, occurring mainly as marine dimethyl-sulfide from the oceans;[7] from Vitousek et al.[8] we learn that thirty to fifty percent of the land surface has been transformed by human action; more nitrogen is now fixed synthetically and applied as fertilizers in agriculture than fixed naturally in all terrestrial ecosystems; the escape into the atmosphere of NO [nitric oxide] from fossil fuel and biomass combustion likewise is larger than the natural inputs, giving rise to photochemical ozone ("smog") formation in extensive regions of the world; more than half of all accessible fresh water is used by mankind; human activity has increased the species extinction rate by one thousand to ten-thousand-fold in the tropical rain forests[9] and several climatically important "greenhouse" gases have substantially increased in the atmosphere: $CO_2$ by more than thirty percent and $CH_4$ [methane] by even more than 100 percent. Furthermore, mankind releases many toxic substances in the environment and even some, the chlorofluorocarbon gases, which are not toxic at all, but which nevertheless have led to the Antarctic "ozone hole" and which would have destroyed much of the ozone layer if no international regulatory measures to end their production had been taken. Coastal wetlands are also affected by humans, having resulted in the loss of fifty percent of the world's mangroves. Finally, mechanized human predation ("fisheries") removes more than twenty-five percent of the primary production of the oceans in the upwelling regions and thirty-five percent in the temperate continental shelf regions.[10] Anthropogenic effects are also well illustrated by the history of biotic communities that leave remains in lake sediments. The effects documented include modification of the geochemical cycle in large freshwater systems and occur in systems remote from primary sources.[11-13]

Considering these and many other major and still growing impacts of human activities on earth and atmosphere, and at all scales, including global, it seems to us more than appropriate to emphasize the central role of mankind in geology and ecology by proposing to use the term "anthropocene" for the current geological epoch. The impacts of current human activities will continue over long periods. According to a study by Berger and Loutre,[14] because of the anthropogenic emissions of $CO_2$ climate may depart significantly from natural behavior over the next 50,000 years.

To assign a more specific date to the onset of the "anthropocene" seems somewhat arbitrary, but we propose the latter part of the eighteenth century, although we are aware that alternative proposals can be made (some may even want to include the entire Holocene). However, we choose this date because, during the past two centuries, the global effects of human activities have become clearly noticeable. This is the period when data retrieved from glacial ice cores show the beginning of a growth in the atmospheric concentrations of several "greenhouse gases," in particular $CO_2$ and $CH_4$.[7] Such a starting date also coincides with James Watt's invention of the steam engine in 1784. About at that time, biotic assemblages in most lakes began to show large changes.[11-13]

Without major catastrophes like an enormous volcanic eruption, an unexpected epidemic, a large-scale nuclear war, an asteroid impact, a new ice age, or continued plundering of Earth's resources by partially still-primitive technology (the last four dangers can, however, be prevented in a real functioning noösphere) mankind will remain a major geological force for many millennia, maybe millions of years, to come. To develop a worldwide accepted strategy leading to sustainability of ecosystems against human-induced stresses will be one of the great future tasks of mankind, requiring intensive research efforts and wise application of the knowledge thus acquired in the noösphere, better known as knowledge or information society. An exciting, but also difficult and daunting task lies ahead of the global research and engineering community: to guide mankind towards global, sustainable, environmental management.[15]

*We thank the many colleagues, especially the members of the IGBP Scientific Committee, for their encouraging correspondence and advice.*

*For casual observers, one of the harder parts of the climate debate was figuring out what most scientists really thought—the average newspaper account tended to cite one proponent and one foe of the idea that global warming was real. Naomi Oreskes, who has taught at Harvard, Dartmouth, and NYU, decided to investigate: this analysis, which appeared in* Science *in December 2004, looked at 928 recent scientific papers and concluded that virtually all accepted the idea that climate change was a real phenomenon, not a hoax. The paper was widely cited; Al Gore referred to it in* An Inconvenient Truth. *She followed up the paper with a widely-read book,* Merchants of Doubt, *which demonstrated that many climate skeptics had made skepticism their (lucrative) life's work, also challenging scientific consensus on topics like the link between cigarettes and cancer.*

# The Scientific Consensus on Climate Change

*Naomi Oreskes*
*2004*

Policy-makers and the media, particularly in the United States, frequently assert that climate science is highly uncertain. Some have used this as an argument against adopting strong measures to reduce greenhouse gas emissions. For example, while discussing a major U.S. Environmental Protection Agency report on the risks of climate change, then-EPA administrator Christine Whitman argued, "As [the report] went through review, there was less consensus on the science and conclusions on climate change."[1] Some corporations whose revenues might be adversely affected by controls on carbon dioxide emissions have also alleged major uncertainties in the science.[2] Such statements suggest that there might be substantive disagreement in the scientific community about the reality of anthropogenic climate change. This is not the case.

The scientific consensus is clearly expressed in the reports of the Intergovernmental Panel on Climate Change (IPCC). Created in 1988 by the World Meteorological Organization and the United Nations Environmental Programme, IPCC's purpose is to evaluate the state of climate science as a basis for informed policy action, primarily on the basis of peer-reviewed and published scientific literature.[3] In its most recent assessment, IPCC states unequivocally that the consensus of scientific opinion is that Earth's climate is being affected by human activities: "Human activities ... are modifying the concentration of atmospheric constituents ... that absorb or scatter radiant energy.... [M]ost of the observed warming over the last 50 years is likely to have been due to the increase in greenhouse gas concentrations."[4]

IPCC is not alone in its conclusions. In recent years, all major scientific bodies in the United States whose members' expertise bears directly on the matter have issued similar statements. For example, the National Academy of Sciences report, *Climate Change Science: An Analysis of Some Key Questions*, begins: "Greenhouse gases are accumulating in Earth's atmosphere as a result of human activities, causing surface air temperatures and subsurface ocean temperatures to rise."[5] The report explicitly asks whether the IPCC assessment is a fair summary of professional scientific thinking, and answers yes: "The IPCC's conclusion that most of the observed warming of the last 50 years is likely to have been due to the increase in greenhouse gas concentrations accurately reflects the current thinking of the scientific community on this issue."[6]

Others agree. The American Meteorological Society,[7] the American Geophysical Union,[8] and the American Association for the Advancement of Science (AAAS) all have issued statements in recent years concluding that the evidence for human modification of climate is compelling.[9]

The drafting of such reports and statements involves many opportunities for comment, criticism, and revision, and it is not likely that they would diverge greatly from the opinions of the societies' members. Nevertheless, they might downplay legitimate dissenting opinions. That hypothesis was tested by analyzing 928 abstracts, published in refereed scientific journals between 1993 and 2003, and listed in the ISI database with the keywords "climate change."[10]

The 928 papers were divided into six categories: explicit endorsement of the consensus position, evaluation of impacts, mitigation proposals, methods, paleoclimate analysis, and rejection of the consensus position. Of all the papers, 75% fell into the first three categories, either explicitly or implicitly accepting the consensus view; 25% dealt with methods or paleoclimate, taking no position on current anthropogenic climate change. Remarkably, none of the papers disagreed with the consensus position.

Admittedly, authors evaluating impacts, developing methods, or studying paleoclimatic change might believe that current climate change is natural. However, none of these papers argued that point.

This analysis shows that scientists publishing in the peer-reviewed literature agree with IPCC, the National Academy of Sciences, and the public statements of their professional societies. Politicians, economists, journalists, and others may have the impression of confusion, disagreement, or discord among climate scientists, but that impression is incorrect.

The scientific consensus might, of course, be wrong. If the history of science teaches anything, it is humility, and no one can be faulted for failing to act on what is not known. But our grandchildren will surely blame us if they find that we understood the reality of anthropogenic climate change and failed to do anything about it.

Many details about climate interactions are not well understood, and there are ample grounds for continued research to provide a better basis for understanding climate dynamics. The question of what to do about climate change is also still open. But there is a scientific consensus on the reality of anthropogenic climate change. Climate scientists have repeatedly tried to make this clear. It is time for the rest of us to listen.

How much carbon is too much? This is the last truly important unanswered question about climate change. In the preindustrial era, $CO_2$ concentrations had hovered around 275 parts per million for thousands of years, producing the climatic stability that underwrote civilization. Post-1988, many analysts seized on a doubled $CO_2$—550 ppm—as the red line, mostly because it was easy to model on a computer. By 2000, some of the big environmental NGOs were advocating a lower target, 450 ppm. But by the summer of 2007, with atmospheric $CO_2$ only nudging 390 parts per million and the Arctic rapidly melting, that analysis was called into question. I asked James Hansen if we knew enough to set a value, and Hansen and his team responded in January 2008 with a landmark paper identifying the most important number on earth: 350 parts per million $CO_2$. Note especially the precise wording they use: values for carbon in excess of that line are "not compatible with the planet on which civilization developed or to which life on earth is adapted." The paper gave birth to the 350.org movement, which has become the largest global grassroots climate campaign.

# Target Atmospheric CO$_2$: Where Should Humanity Aim?

James Hansen, Makiko Sato, Pushker Kharecha,
David Beerling, Robert Berner, Valerie Masson-Delmotte,
Mark Pagani, Maureen Raymo, Dana L. Royer, and
James C. Zachos
2008

*Abstract*

Paleoclimate data show that climate sensitivity is approximately three degrees Celsius for doubled $CO_2$, including only fast feedback processes. Equilibrium sensitivity, including slower surface albedo feedbacks, is approximately six degrees Celsius for doubled $CO_2$ for the range of climate states between glacial conditions and ice-free Antarctica. Decreasing $CO_2$ was the main cause of a cooling trend that began fifty million years ago, the planet being nearly ice-free until $CO_2$ fell to 450 plus or minus 100 parts per million; barring prompt policy changes, that critical level will be passed, in the opposite direction, within decades. If humanity wishes to preserve a planet similar to that on which civilization developed and to which life on Earth is adapted, paleoclimate evidence and ongoing climate change suggest that $CO_2$ will need to be reduced from its current 385 parts per million to at most 350 parts per million, but likely less than that. The largest uncertainty in the target arises from possible changes of non-$CO_2$ forcings. An initial 350 parts per million $CO_2$ target may be achievable by phasing out coal use except where $CO_2$ is captured and adopting agricultural and forestry practices that sequester carbon. If the present overshoot of this target $CO_2$ is not brief, there is a possibility of seeding irreversible catastrophic effects.

## Introduction

Human activities are altering Earth's atmospheric composition. Concern about global warming due to long-lived human-made greenhouse gases (GHGs) led to the United Nations Framework Convention on Climate Change,[1] with the objective of stabilizing GHGs in the atmosphere at a level preventing "dangerous anthropogenic interference with the climate system."

The Intergovernmental Panel on Climate Change [IPCC][2] and others[3] used several "reasons for concern" to estimate that global warming of more than two to three degrees Celsius may be dangerous. The European Union adopted two degrees above pre-industrial global temperature as a goal to limit human-made warming.[4] Hansen et al.[5] argued for a limit of one degree Celsius global warming (relative to 2000, 1.7 degrees relative to pre-industrial time), aiming to avoid practically irreversible ice-sheet and species loss. This one-degree limit, with nominal climate sensitivity of 0.75 degrees Celsius per watt per square meter and plausible control of other GHGs,[6] implies maximum $CO_2$ approximately 450 parts per million.[5]

Our current analysis suggests that humanity must aim for an even lower level of GHGs. Paleoclimate data and ongoing global changes indicate that "slow" climate feedback processes not included in most climate models, such as ice-sheet disintegration, vegetation migration, and GHG release from soils, tundra or ocean sediments, may begin to come into play on time scales as short as centuries or less.[7] Rapid ongoing climate changes and the realization that Earth is out of energy balance, implying that more warming is "in the pipeline",[8] add urgency to investigation of the dangerous level of GHGs.

A probabilistic analysis[9] concluded that the long-term $CO_2$ limit is in the range 300–500 parts per million for twenty-five percent risk tolerance, depending on climate sensitivity and non-$CO_2$ forcings. Stabilizing atmospheric $CO_2$ and climate requires that net $CO_2$ emissions approach zero, because of the long lifetime of $CO_2$.[10, 11]

We use paleoclimate data to show that long-term climate has high sensitivity to climate forcings and that the present global mean $CO_2$, 385 parts per million, is already in the dangerous zone. Despite rapid current $CO_2$ growth, approximately two parts per million per year, we show that it is conceivable to reduce $CO_2$ this century to less than the current amount, but only via prompt policy changes.

[. . .]

## Summary

Humanity today, collectively, must face the uncomfortable fact that industrial civilization itself has become the principal driver of global climate. If we stay our present course, using fossil fuels to feed a growing appetite for energy-intensive lifestyles, we will soon leave the climate of the Holocene, the world of prior human history. The eventual response to doubling pre-industrial atmospheric $CO_2$ likely would be a nearly ice-free planet, preceded by a period of chaotic change with continually changing shorelines.

Humanity's task of moderating human-caused global climate change is urgent. Ocean and ice-sheet inertias provide a buffer delaying full response by centuries, but there is a danger that

Fig. 7 (a)   Temperature Anomaly (°C): Seasonal Resolution

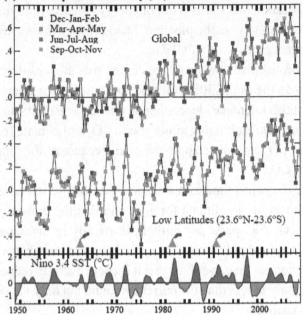

(b) Global-Ocean Mean SST Anomaly (°C): Monthly Resolution

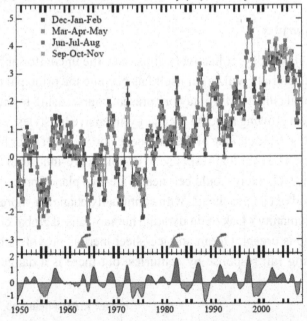

human-made forcings could drive the climate system beyond tipping points such that change proceeds out of our control. The time available to reduce the human-made forcing is uncertain, because models of the global system and critical components such as ice sheets are inadequate. However, climate response time is surely less than the atmospheric lifetime of the human-caused perturbation of $CO_2$. Thus remaining fossil-fuel reserves should not be exploited without a plan for retrieval and disposal of resulting atmospheric $CO_2$.

Paleoclimate evidence and ongoing global changes imply that today's $CO_2$, about 385 parts per million, is already too high to maintain the climate to which humanity, wildlife, and the rest of the biosphere are adapted. The realization that we must reduce the current $CO_2$ amount has a bright side: effects that had begun to seem inevitable, including impacts of ocean acidification, loss of fresh water supplies, and shifting of climatic zones, may be averted by the necessity of finding an energy course beyond fossil fuels sooner than would otherwise have occurred.

We suggest an initial objective of reducing atmospheric $CO_2$ to 350 parts per million, with the target to be adjusted as scientific understanding and empirical evidence of climate effects accumulate. Although a case already could be made that the eventual target probably needs to be lower, the 350 parts per million target is sufficient to qualitatively change the discussion and drive fundamental changes in energy policy. Limited opportunities for reduction of non-$CO_2$ human-caused forcings are important to

Fig. (7). (a) Seasonal-mean global and low-latitude surface temperature anomalies relative to 1951–1980, an update of [92]; (b) global-ocean-mean sea surface temperature anomaly at monthly resolution. The Nino 3.4 Index, the temperature anomaly (twelve-month running mean) in a small part of the tropical Pacific Ocean [93], is a measure of ENSO, a basin-wide sloshing of the tropical Pacific Ocean [94]. Green triangles show major volcanic eruptions.

pursue but do not alter the initial 350 parts per million $CO_2$ target. This target must be pursued on a timescale of decades, as paleoclimate and ongoing changes, and the ocean response time, suggest that it would be foolhardy to allow $CO_2$ to stay in the dangerous zone for centuries.

A practical global strategy almost surely requires a rising global price on $CO_2$ emissions and phase-out of coal use except for cases where the $CO_2$ is captured and sequestered. The carbon price should eliminate use of unconventional fossil fuels, unless, as is unlikely, the $CO_2$ can be captured. A reward system for improved agricultural and forestry practices that sequester carbon could remove the current $CO_2$ overshoot. With simultaneous policies to reduce non-$CO_2$ greenhouse gases, it appears still feasible to avert catastrophic climate change.

Present policies, with continued construction of coal-fired power plants without $CO_2$ capture, suggest that decision-makers do not appreciate the gravity of the situation. We must begin to move now toward the era beyond fossil fuels. Continued growth of greenhouse-gas emissions, for just another decade, practically eliminates the possibility of near-term return of atmospheric composition beneath the tipping level for catastrophic effects.

The most difficult task, phase-out over the next twenty to twenty-five years of coal use that does not capture $CO_2$, is Herculean, yet feasible when compared with the efforts that went into World War II. The stakes, for all life on the planet, surpass those of any previous crisis. The greatest danger is continued ignorance and denial, which could make tragic consequences unavoidable.

*If 1958, 1988, and 1995 are crucial dates in the story of global warming, 2011 may prove just as important. In August 2010, scientists reported that we'd just come through the warmest spring on record, the warmest year, and the warmest decade. But there was no longer anything abstract about those numbers, not in the face of some of the greatest heat waves and flooding the world has ever seen. Dr. Jeff Masters, one of the world's most widely read weather bloggers (he is the founder of the Weather Underground website) offered a candid assessment one day in mid-August. You can't say any particular event "is" global warming—but you can say we're loading the dice.*

# Causes of the Russian Heat Wave and Pakistani Floods
## from Dr. Jeff Masters' WunderBlog

Jeff Masters
*2:56 PM GMT on August 13, 2010*

The Great Russian Heat Wave of 2010 is one of the most intense, widespread, and long-lasting heat waves in world history. Only the European heat wave of 2003, which killed 35,000–50,000 people, and the incredible North American heat wave of July 1936, which set all-time extreme highest temperature records in fifteen U.S. states, can compare. All of these heat waves were caused by a highly unusual kink in the jet stream that remained locked in place for over a month. The jet stream is an upper-level river of air, between the altitudes of about 30,000–40,000 feet (10,000–12,000 meters). In July over Europe and Asia, the jet stream has two branches: a strong southern "subtropical" jet that blows across southern Europe, and a weaker "polar" jet that blows across northern Europe. The polar jet stream carries along the extratropical cyclones (lows) that bring the midlatitudes most of their precipitation. The polar jet stream also acts as the boundary between cold, Arctic air, and warm tropical air. If the polar jet stream shifts to the north of its usual location, areas just to its south will be much hotter and drier than normal. In July 2010, a remarkably strong polar jet stream developed over northern Europe. This jet curved far to the north of Moscow,

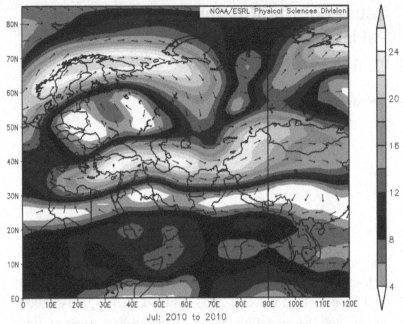

NCEP/NCAR Reanalysis
300mb Vector Wind (m/s) Composite Mean

Jul: 2010 to 2010

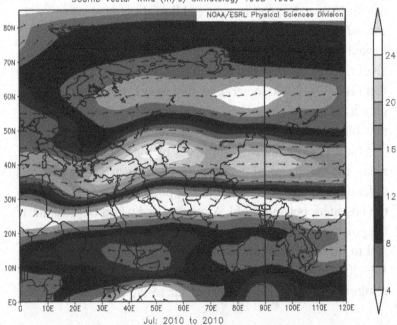

NCEP/NCAR Reanalysis
300mb Vector Wind (m/s) Climatology 1968–1996

Jul: 2010 to 2010

then plunged southwards towards Pakistan. This allowed hot air to surge northwards over most of European Russia, and prevented rain-bearing low pressure systems from traveling over the region. These rain-bearing low pressure systems passed far to the north of European Russia, then dove unusually far to the south, into northern Pakistan. The heavy rains from these lows combined with Pakistan's usual summer monsoon rains to trigger Pakistan's most devastating floods in history.

## What caused this unusual jet stream pattern?

The unusual jet stream pattern that led to the 2010 Russian heat wave and Pakistani floods began during the last week of June, and remained locked in place all of July and for the first half of August. Long-lived "blocking" episodes like this are usually caused by unusual sea surface temperature patterns, according to recent research done using climate models. For example, Feudale and Shukla (2010) found that during the summer of 2003, exceptionally high sea surface temperatures [SSTs] of four degrees Celsius (seven degrees Fahrenheit) above average over the Mediterranean Sea, combined with unusually warm SSTs in the northern portion of the North Atlantic Ocean near the Arctic, combined to shift the jet stream to the north over Western Europe and create the heat wave of 2003. I expect that the current SST pattern over the ocean regions surrounding Europe played a key

---

Winds of the jet stream at an altitude of 300 millibars (roughly 30,000 feet high). Top: Average July winds from the period 1968–1996 show that a two-branch jet stream typically occurs over Europe and Asia—a northern "polar" jet stream, and a more southerly "subtropical" jet stream. Bottom: the jet stream pattern in July 2010 was highly unusual, with a very strong polar jet looping far to the north of Russia, then diving southwards towards Pakistan. Image credit: NOAA/ESRL.

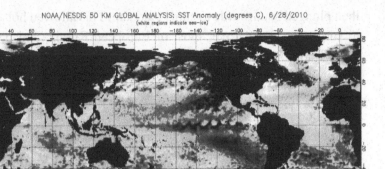

NOAA/NESDIS 50 KM GLOBAL ANALYSIS: SST Anomaly (degrees C), 6/28/2010
(white regions indicate sea-ice)

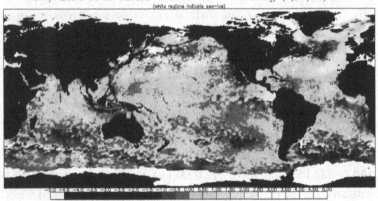

NOAA/NESDIS 50 KM GLOBAL ANALYSIS: SST – Climatology (C), 6/28/2003
(white regions indicate sea-ice)

A comparison of the departure of sea surface temperature (SST) from average just prior to the start of the great European heat waves of 2003 and 2010. Temperatures in the Mediterranean Sea were up to 4 degrees Celsius above average in 2003, which has been implicated as a major cause of the Western European heat wave of 2003. Image credit: NOAA/NESDIS.

role in shifting the jet stream to create the heat wave of 2010. Note that the SST anomaly pattern is quite different this year compared to 2003, which may be why this year's heat wave hit Eastern Europe, and the 2003 heat wave hit Western Europe. Human-caused climate change also may have played a role; using climate models, Stott et al. (2004) found it very likely (a greater than ninety percent chance) that human-caused climate change has at least doubled the risk of severe heat waves like the great 2003 European heat wave.

# PART II
## POLITICS

*Al Gore is the key politician of the global warming era. As a Tennessee senator, he held a number of early hearings on climate change (giving James Hansen, among others, an important platform). But he really adopted the issue as his own in the early 1990s, at a moment when a swelling demand for action seemed to make it a political winner. His book,* Earth in the Balance, *came out in 1992, the same year he won the vice presidency. It's a clarion call for strong action, and includes these pages that famously demand that environmental preservation be the "organizing principle" of the planet's politics in the decades ahead.*

# from *Earth in the Balance*

Al Gore
2000

Consider Cancer Alley in the Lower Mississippi River Valley between Baton Rouge and New Orleans, where more than a quarter of America's chemicals are produced and where some of the highest cancer rates in the nation are found. Pat Bryant, an African-American political activist who got his start in the early 1980s by organizing public housing tenants in St. Charles Parish, shifted his attention to the constant respiratory and eye problems of the children who lived near the Union Carbide and Monsanto complexes. In Bryant's view—a view shared by many others—Cancer Alley was made possible by ethnic discrimination and political powerlessness.

I met Bryant in Atlanta at the Southern Environmental Assembly, a gathering of mostly white people. As he said later, "A lot of the environmentalists were middle class. We all speak English, but what we say doesn't always mean the same thing. We must put aside foolish customs that divide us and work together, at least for the sake of our children." True to this vision, Bryant organized a coalition of environmental and labor groups to create the Louisiana Toxics Project, which contributed to the passage of the state's first air quality law in 1989.

But the coalition wasn't finished, and Bryant's view of the problem extended beyond Cancer Alley. The following year, during the Senate's consideration of the Clean Air Act, Bryant and one of the national groups linked with his project brought a glaring loophole to my attention; it would have allowed companies emitting toxic air pollutants (the most deadly class of air pollution) to avoid the tougher emissions standards by buying up neighborhoods downwind of their facilities and creating what environmentalists call "dead zones," large areas devoid of people and inevitably bordering poor neighborhoods, whose property values would drop. Of course, whenever the wind shifted, the toxic pollutants that were supposed to fall in the dead zone would fall somewhere else—most often on impoverished black families. The national coalition was instrumental in passing the amendment to close the loophole.

Bryant's perspective is especially important because of continuing fears on the part of some activists who work with the poor and oppressed that the environmental movement will divert attention from their priorities. As Bryant puts it, "The environment is the number-one problem in this country. As an African American, my hope and aspiration to be free are greatly dimmed by the prospect of environmental destruction. If we're going to make great strides on this problem, we're going to have to build African American–European American coalitions."

Sometimes, of course, the NIMBY phenomenon raises difficult questions about how and where to locate unpopular facilities. Indeed, among the most highly charged and passionately argued political issues today are proposals to place new landfills or garbage dumps in areas where those nearby feel threatened. But I have found that when the merits of a proposal really do make sense, those who are fighting it often temper their opposition, or at least find it harder to attract support outside their immediate area. More commonly, advocates of a facility that raises serious environmental questions try to distract attention from the real issues by accusing their opponents of adopting a knee-jerk NIMBY approach. And while it's true that people are sometimes too self-interested in their opposition to these issues, in my opinion the NIMBY syndrome is the beginning of a healthy trend. In fact, I am convinced that political support for measures to protect the global atmosphere will one day surge when the meaning of the word "backyard" expands to encompass each person's share of the air we all breathe.

The impetus for that change will come from the forefront of science and from the work of scientists like Dr. Sherwood Rowland, who in 1974 discovered a dramatic change in the chemical composition of our atmosphere. Concentrations of chlorine have increased enormously throughout the world because of the widespread use of chlorofluorocarbons (CFCs). But when he and Dr. Mario Molina, both of the University of California, Irvine, announced their disturbing discovery, Rowland suffered a form of scientific persecution. Suddenly he was no longer invited to address as many scientific meetings; in at least two cases, companies profiting from the suspect chemicals threatened to withhold funding for conferences if Rowland was on the program. But Sherwood Rowland has a keenly developed sense of right and wrong; he decided to fight, and he has been fighting for more than seventeen years now. With his wife, Joanne, he has traveled to conferences and symposia in every part of the world, argued his case, and patiently taken on all comers.

In large part because of the steadfast work of Sherwood Rowland and such colleagues as Mario Molina and Robert Watson of NASA, the world was ready to listen when the ozone hole caused by CFCs suddenly appeared over Antarctica in 1987. Dr. Susan Solomon led an emergency scientific expedition to the South Pole and confirmed what Rowland had hypothesized. Many countries have finally begun taking action, yet even now, when the evidence against CFCs is overwhelming, these life- threatening compounds are still being released into the atmosphere, and some countries still refuse to join the global effort to ban them.

Some members of the resistance have taken the fight for the environment from the scientific journals and symposia to their own backyards and from there to corporate boardrooms and the halls of Congress. One remarkable woman, Lynda Draper, joined the fight in her own kitchen. I learned about her courageous battle in early 1989, when she came to my office for help soon after discovering that General Electric (GE) was planning to release a huge quantity of CFCs into the atmosphere; in fact, they had already begun doing so. As she told me the story (later confirmed by GE), a repairman had knocked on her door in Ellicott City, Maryland, and said that her relatively new refrigerator had a defective compressor that needed to be replaced. Indeed, some GE officials felt they were demonstrating foresight and responsiveness in organizing the largest recall program in the history of the industry; they intended to replace an estimated one to two million compressors that might otherwise fail, leaving some customers with spoiled food.

As Draper told it, the repairman came into her kitchen and inspected her refrigerator. "Then he asked me to open the window. I didn't know why he wanted me to do that, but I did. Then all of a sudden I heard this loud whoosh!" Draper, who had worked for environmental groups, immediately realized what was happening: the CFCs from the old compressor were being vented from her refrigerator through her window right into the atmosphere. Horrified, she protested to the repairman. When he explained that it was only a few ounces, she didn't buy it. She went to work finding out how many refrigerators were involved in the recall program and multiplied the total by the number of ounces of CFCs in each; she determined that at least 125 and perhaps as many as 312 tons of CFCs would enter the atmosphere over the course of the recall program. She was determined to stop the company, but the challenge she accepted was steeped in irony: both her father and her grandfather had been long-standing GE employees, and her husband had worked for the company for ten years. At first, Draper took the most obvious course of action—she called up the company to tell them what they were doing and why it was wrong. When the company responded that the amounts were too small to worry about, she decided to complain to local and state officials and finally the EPA. Even then, she got nowhere. By the time she arrived at my office, she had contacted the Public Interest Research Group and had made plans for a press conference to call for a nationwide boycott of all GE products.

In response to Draper's persistent efforts, the company completely changed its corporate policy on CFCs and became the industry leader in CFC reductions, setting standards its competitors are still trying to match. GE developed special equipment for scavenging CFCs instead of releasing them and used this equipment to remove CFCs from other parts of the environment as a way of offsetting what they had released during the recall program. The proposed boycott was never called, and Draper, who started as a PTA volunteer, now works full-time to save the environment. "I intend to keep fighting," she says. "If more people were fighting, we'd make more progress."

Sherwood Rowland and Lynda Draper are, in effect, comrades in arms in the same struggle. But the struggle isn't just about CFCs. Ultimately, it's about the entire relationship between human civilization and the global environment. Slowly, people in all walks of life are coming to understand the enormity of the problem; slowly we are awakening to the strategic threats now posed by our rapidly expanding civilization. Although the resistance is growing, becoming more sophisticated, and scoring some dramatic successes, the larger war to save the earth is being lost. That will change only when the rest of humankind, drawing on the lessons learned by these pioneers and inspired by their courage and sacrifice, finally organizes an all-out response to this unprecedented threat.

Again, we must not forget the lessons of World War II. The Resistance slowed the advance of fascism and scored important victories, but fascism continued its relentless march to domination until the rest of the world finally awoke and made the defeat of fascism its central organizing principle from 1941 through 1945. But too many ignored the early warnings; in June 1936, for instance, Haile Selassie, the emperor of Ethiopia, addressed the world through the League of Nations. His was the first country invaded by the Axis, and in describing the atrocities committed by Mussolini's forces, including the use of poison gas, he said, "Soldiers, women, children, cattle, rivers, lakes, and pastures were drenched continually with this deadly rain. In order to kill off systematically all living creatures, in order the more surely to poison waters and pastures, the Italian command made its aircraft pass over and over again." Selassie said that he wanted both to describe the atrocities against his people and to make it plain that the rest of the world would soon face the same aggression. He came, he said, to "give Europe a warning of the doom that awaits it if it should bow before the accomplished fact…. God and history will remember your judgment."

The world is once again at a critical juncture. A relentless advance is again claiming victims throughout the world, and again courageous men and women are standing in the path of destruction and calling upon the rest of the world to help stop the invasions. But this time we are invading ourselves and attacking the ecological system of which we are a part. As a result, we now face the prospect of a kind of global civil war between those who refuse to consider the consequences of civilization's relentless advance and those who refuse to be silent partners in the destruction. More and more people of conscience are joining the effort to resist, but the time has come to make this struggle the central organizing principle of world civilization. We have had a warning of the fate that awaits if we "bow before the accomplished fact." God and history will remember our judgment.

*In the years after the 1992 UN Conference on Environment and Development in Rio de Janeiro, climate science continued to strengthen the case for action. With Al Gore as vice president and the United States topping the list of global carbon dioxide emitters, Washington was the most obvious place for that action. But Congress showed no sign of progress. Ross Gelbspan exposed the root of this apathy with his book The Heat Is On, the first full-scale attempt to show that the fossil fuel industry was organizing and funding a serious resistance to the science of global warming. He detailed how a small coterie of well-funded skeptics was receiving outsize press attention and confusing the public into believing that scientists were divided on the issue. Gelbspan, a veteran reporter who had edited a Pulitzer Prize–winning series at the Boston Globe, was not a longstanding environmentalist—at least initially, it was the corruption inherent in the industry effort that outraged him. But he has gone on to be one of the most powerful and uncompromising journalistic voices for real change. Meanwhile, many of the industry-funded contrarians he describes carry on with their campaign to deny the science.*

# The Battle for Control of Reality
from *The Heat Is On*

Ross Gelbspan
*1998*

The financial resources available to the oil and coal lobbies are almost without limit. They can buy Congress. In fact, long before the climate issue surfaced, they already had.

They can buy media access. Not just the Mobil ads, eye-catchingly conspicuous on the *New York Times* op-ed page, but also access to editorial boards, TV producers, and every relevant reporter in the country.

Over the last six years the coal and oil industries have spent millions of dollars to wage a propaganda campaign to downplay the threat of climate change. Much of that money has gone to amplifying the views of about a half-dozen dissenting research-ers, giving them a platform and a level of credibility in the public arena that is grossly out of proportion to their influence in the scientific community.

The campaign to keep the climate change off the public agen-da involves more than the undisclosed funding of these "green-house skeptics." In their efforts to challenge the consensus scien-tific view about the escalating turmoil of the global atmosphere, the public relations apparatus of the oil and coal industries has publicized weather reports, and funded and distributed books and videos and self-proclaimed journals of science that are dismissed by the vast majority of mainstream scientists.

John Stauber is the editor of *PR Watch*, a newsletter that monitors the $10-billion-a-year public relations industry. In his book *Toxic Sludge is Good for You*, he quotes one public relations executive as saying: "Persuasion by its definition is subtle. The best PR ends up looking like news. You never know when a PR agency is being effective. You'll just find your views slowly shifting." The quote applies, with uncanny accuracy, to the campaign to carefully manufacture public confusion about global climate change. Big oil and big coal have successfully created the general perception that climate scientists are sharply divided over the extent and the likely impacts of climate change—and even over whether it is taking place at all.

The Information Council on the Environment (ICE) was the creation of a group of utility and coal companies. In 1991, using the ICE, the coal industry launched a blatantly misleading campaign on climate change that had been designed by a public relations firm. This public relations firm clearly stated that the aim of the campaign was to "reposition global warming as theory rather than fact." Its plan specified that three of the so-called greenhouse skeptics—Robert Balling, Pat Michaels, and S. Fred Singer—should be placed in broadcast appearances, op-ed pages, and newspaper interviews.

With all the sophistication of modern marketing techniques, the ICE campaign targeted "older, less-educated men" and "young, low-income women" in electoral districts that get their electricity from coal and that, preferably, have a congressperson on the House Energy Committee. The campaign was clever if not accurate. One newspaper advertisement prepared by the ICE, for example, was headlined: "If the earth is getting warmer, why is Minneapolis getting colder?" (Data indicate that the Minneapolis area has actually warmed between 1 and 1.5 degrees Celsius in the last century.)

The ICE campaign had barely gathered momentum when it was exposed by environmentalists, who provided information about it to the media. Several embarrassing news articles led to the ICE's demise. But its case illustrates how a little "repositioning" can generate tremendous public confusion about the state of scientific understanding.

In 1995 I co-authored an article for *The Washington Post* on new research that linked disease outbreaks in various places around the world to changes in local climate. The article was generally well received, but a number of readers wrote to assure me that there was no proven link between industrial emissions and global climate change. Their responses caused me to wonder why so many people believed that the validity of the issue was still so far from certain.

It did not take me long to find the answer.

Ever since climate change took center stage at the 1992 UN Conference on Environment and Development (UNCED) in Rio de Janeiro, Pat Michaels and Robert Balling, together with Sherwood Idso, S. Fred Singer, Richard S. Lindzen, and a few other high-profile greenhouse skeptics have proven extraordinarily adept at draining the issue of all sense of crisis. They have made frequent pronouncements on radio and television programs, including a number of appearances by some of them on the Rush Limbaugh show, their interviews, columns, and letters have appeared in newspapers ranging from local weeklies to

*The Washington Post* and *The Wall Street Journal*. In the process they have helped create a broad public belief that the question of climate change is hopelessly mired in unknowns.

If the climate skeptics have succeeded in confusing the general public, their influence on decision-makers has been, if anything, even more effective.

Their testimony contributed to the defeat of proposals in California and Colorado to increase electricity rates to reduce the amount of greenhouse gases produced by oil- and coal-burning power plants. (A similar initiative was approved recently by the state of Minnesota.) Congressional conservatives have used the testimony of the skeptics to justify cutting the climate research budgets and to discredit the scientific findings of the IPCC [Intergovernmental Panel on Climate Change].

The origins of the prominence of most of these greenhouse skeptics are spelled out, remarkably enough, in several annual reports of the $400 million coal giant, the Western Fuels Association. In its 1994 annual report, Western Fuels declared quite candidly that "There has been a close to universal impulse in the [fossil fuel] trade association community here in Washington to concede the scientific premise of global warming ...while arguing over policy prescriptions that would be the least disruptive to our economy.... We have disagreed, and do disagree, with this strategy."

Western Fuels elaborated on its approach in another report.

When [the climate change] controversy first erupted at the peak of summer in 1988, Western Fuels Association decided it was important to take a stand.... [S]cientists were found who are skeptical about much of what seemed generally accepted about the potential for climate change. Among them were [Pat] Michaels, Robert Balling of Arizona State University, and S. Fred Singer of the University of Virginia…. Western Fuels approached Pat Michaels about writing a quarterly publication designed to provide its readers with critical insight concerning the global climatic change and greenhouse effect controversy…. Western Fuels agreed to finance publication and distribution of *World Climate Review* magazine.

In 1991, before the ill-fated ICE campaign had been buried, one of its funders, Western Fuels, spent $250,000 to produce a video. The video was shown extensively in Washington as well as in the capitals of the OPEC nations. Insiders at the Bush White House said it was Chief of Staff John Sununu's favorite movie—he showed it that often. Then Secretary of Energy James Watkins cited it in a conversation during a visit to *The Boston Globe*, where I interviewed him. The video's aim is to persuade policymakers that a warmer, wetter, carbon dioxide–enhanced world would be, contrary to the alarms of environmentalists, a godsend.

Titled *The Greening of Planet Earth*, the video is narrated by Sherwood Idso, an active skeptic, and it features Richard Lindzen as well as a number of botanists and agronomists. (The video was produced by a company headed by Idso's wife; the coal industry, which funded it, has purchased and circulated hundreds of copies of books and publications written by Idso.) In near-evangelical tones, it promises that a new age of agricultural abundance will result from the doubling of the atmospheric concentration of carbon dioxide. It shows plant biologists predicting that yields of soybeans, cotton, wheat, and other crops will increase by thirty to sixty percent—enough to feed and clothe the earth's expanding human population. The video portrays a world where vast areas of desert are replaced by grasslands, where today's grass- and scrublands are transformed by a new cover of bushes and trees, and where today's diminishing forests are replenished by new growth as a result of a nurturing atmosphere of enhanced carbon dioxide.

Unfortunately, it overlooks the bugs.

Insects are extremely sensitive to changes in temperature. According to a panel of the World Health and World Meteorological organizations and the UN Environmental Programme (UNEP), even a minor elevation in temperature would trigger an explosion in the planet's insect population. A slight warming, the panel suggests, could result in insect-related crop damage, leading to a significant disruption in the food supply. The spread of insect-borne diseases could surge. The panel, which examines interactions between global climate and biological systems, notes that the *Aedes aegypti* mosquito, which spreads dengue fever and yellow fever, has traditionally been unable to survive at altitudes higher than 1,000 meters because of colder temperatures there. But with recent warming trends, those mosquitoes have now been reported at 1,240 meters in Costa Rica and at 2,200 meters in Colombia. Malaria-bearing mosquitoes, too, have moved to higher elevations in central Africa, Asia, and parts of Latin America, triggering new outbreaks of the disease.

It appears that while the coal-funded video extolled the benefits of global warming, it neglected to tell people what the warming-induced infestation of termites in New Orleans may be trying to tell them now.

(In fairness, many agricultural scientists also acknowledge

some short-term benefits of enhanced $CO_2$. In the short term, it may indeed increase yields and growth rates of food crops in the mid-northern latitudes—to the benefit of U.S., Canadian, and Russian agriculture. But other effects are not so positive. The initial increases in crop growth and food yield, many scientists fear, will soon flatten, and a long-term diet of concentrated carbon dioxide will weaken the plants, making them less robust.

(More importantly, enhanced $CO_2$ would be devastating to crop growth in the poor areas of the world—the midtropical regions. In many of these areas, population growth is already stressing the food-growing capability of the land. Increased $CO_2$, agricultural researchers say, will force more rapid plant respiration. When that forcing is accompanied by slightly elevated temperatures, the plants will stop growing and their yield will dwindle. A significant enhancement of $CO_2$ in the earth's most heavily populated middle-warm belt will accelerate rates of malnutrition, disease, and starvation.)

Western Fuels is not alone in its efforts to veil the reality of climate change. On the eve of the March 1995 round of international climate negotiations (which had been set up by the UNCED conference in Rio), another industry lobbying group, the Global Climate Coalition (GCC), disseminated a report issued by the private weather-forecasting firm Accu-Weather. The report declared there has been no increase in severe weather events. A GCC press release that accompanied the report noted that "Accu-Weather experts say there is no convincing evidence that global weather is becoming more extreme." It quoted an Accu-Weather executive as saying, "Scientific evidence squarely disputes the hypothesis that hurricanes are becoming stronger and more frequent, that tornadoes have increased in number, and that droughts and floods are becoming more common. In fact, the data show that...temperature and precipitation extremes are no more common now than they were fifty to one hundred years ago."

The report was dismissed by a number of mainstream scientists, who noted that it contradicted the findings of a team of researchers from the NOAA National Climatic Data Center earlier that year. The NOAA team had demonstrated that the increase in severe weather events had been fueled by atmospheric warming. Several scientists took issue with the Accu-Weather report's methodology, pointing out that its weather and temperature readings had been taken from only three data points—Augusta, Georgia, State College, Pennsylvania, and Des Moines, Iowa—hardly a broad-based sample.

Equally telling, the Accu-Weather report flies in the face of insurance industry figures that show that annual weather-related disaster claims have increased sixfold since the 1980s, from $5 billion to $30 billion in the first half of the 1990s.

In May 1995 Judge Allan Klein, who sits on the Minnesota Public Utilities Commission, was charged with the responsibility of determining the environmental costs of the burning of coal by Minnesota power plants. In his administrative courtroom in St. Paul, Judge Klein heard testimony from four greenhouse skeptics who had been hired as expert witnesses not only on behalf of Western Fuels Association, but of several local utilities and the state of North Dakota, the largest supplier of coal to neighboring Minnesota.

Called to the stand was Richard Lindzen, a professor of meteorology at MIT, who testified that, given current trends, the likeliest increase in atmospheric warming by the middle of the next century would be 0.3 degrees Celsius. Although global emissions of carbon dioxide and other greenhouse gases will actually double by the year 2040, Lindzen, a short, owlish-looking man with a professorial demeanor, an argumentative style, a quick sense of humor, and $2 million in federal research grants in his distinguished thirty-year academic career, believes the impacts, if any, will be negligible.

Also called to testify was Pat Michaels, associate professor of climatology at the University of Virginia, who told Judge Klein that despite the buildup of greenhouse gases, he foresees no increase in the rate of sea level rise—a feared consequence of global warming. The gregarious, engaging Michaels is a frequent commentator on climate issues and the founder and publisher of *World Climate Review* and its successor publication, *World Climate Report*.

Robert Balling, a professor of climatology at Arizona State University, concurred that any potential warming is barely worth consideration. "I would anticipate no more than a small rise in temperature, maybe a degree," he testified in St. Paul. "Whether it will continue to move linearly or continue to move in some other fashion is something that's so speculative, it's almost useless to think about it." Balling, a Marine-trim, boyish-looking forty-three-year-old author of a book on global warming titled *The Heated Debate*, has labeled concerns about climate change "pure media hype."

The power of this apparent certainty should not be underestimated.

Professor Willett Kempton, a senior policy scientist at the University of Delaware, has documented the influence of the greenhouse skeptics in *Environmental Values in American Culture*, a book he coauthored that was funded by the National Science Foundation. An aide to a Republican congressman who supports health-related environmental regulation and endangered species protection told Kempton about a televised presentation he had heard by one skeptic. "It came out pretty clearly...that there is a range of disagreement, and [that] two people with equally impressive credentials can disagree," he said. After hearing several more skeptics, the aide decided that scientifically "There's no mainstream, there's no fringes, there are just people all over the lot." Moreover, a union lobbyist interviewed for the same book said: "Everything stems from the assumption that the earth is getting warmer and the causes are [greenhouse gases].... But when I read opposing articles...they're actually...more persuasive than the others." The skeptics led him to believe that "very little [has been] done to measure changes in temperature of oceans or the air mass above the oceans. And there has not really been the depth of scientific inquiry necessary to say that... is a problem."

The tiny group of dissenting scientists have been given prominent public visibility and congressional influence out of all proportion to their standing in the scientific community on the issue of global warming. They have used this platform to pound widely amplified drumbeats of doubt about climate change. These doubts are repeated in virtually every climate-related story in every newspaper and every TV and radio news outlet in the country.

By keeping the discussion focused on whether there really is a problem, these dozen or so dissidents—contradicting the consensus view held by 2,500 of the world's top climate scientists—have until now prevented discussion about how to address the problem.

The skeptics are virtually unanimous in accusing their mainstream scientific colleagues of exaggerating the magnitude of the climate problem in order to perpetuate their own government research funding.

But that argument cuts both ways.

While testifying in St. Paul, Pat Michaels revealed under oath that he had received more than $165,000 in industry and private funding over the previous five years—funding he had never previously disclosed. Not only did Western Fuels fund both his publications, he disclosed, but it provided a $63,000 grant for his research. Another $49,000 came to Michaels from the German Coal Mining Association. A smaller grant of $15,000 came from the Edison Electric Institute. Michaels also listed a grant of $40,000 from the Western mining company Cyprus Minerals. Questioned by the assistant attorney general about that grant, Michaels responded, "You know, with all due respect, you're going to think I'm not telling the truth. I'm trying to remember directly what came out of the project.... I'm sure we were looking at regional temperatures in some way."

In fact, Cyprus Minerals was at the time the largest single funder of the virulently antienvironmentalist Wise Use movement. The biggest organizational member of that movement was a group called People for the West!, whose largest funder, with at least $100,000 in donations, was Cyprus Minerals. According to the Clearinghouse in Environmental Advocacy and Research, as recently as 1995 Cyprus Minerals' director of governmental affairs was a member of the board of directors of People for the West!.

In interviews, Michaels has insisted that he dissociated himself from the ICE campaign when he learned of what he called its "blatant dishonesty." But he apparently had no qualms about accepting money to publish his own journal, *World Climate Review*, from one of the same coal-industry sources that funded the ICE campaign. (The industry funding of Michaels's publications was first made public by Bud Ward, editor of *Environment Writer*, the newsletter for journalists published by the Environmental Health Center of the National Safety Council. Unfortunately the journalists Ward writes for made little use of the information.) Michaels, for his part, insists that this now-defunct journal, as well as its successor coal-funded publication, *World Climate Report*, which Michaels also edits, are serious journals of climate science.

However, a reading of those publications reveals passages such as this one, written by Michaels in the fall 1994 issue of *World Climate Review*: "The fact is that the artifice of climate-change-as-apocalypse is crumbling faster than Cuba.... There is genuine fear in the environmental community about this one, for the decline and fall of such a prominent issue is sure to horribly maim the credibility of the green movement that espoused it so cheerily."

This is not the language of science, such as one finds in *Science*, *Nature*, or *The Bulletin of the American Meteorological Society*. It is the language of propaganda.

The winter 1993 issue of Michaels's magazine featured a cover photo that appeared to replicate the front page of *The Washington Post*, with the headline "The End Nears Again." Michaels was subsequently forced to apologize to the *Post* for his choice of cover art, which more closely resembled a cover for the *National Lampoon* than one for a journal of science. In the winter 1993 issue, he wrote of mainstream scientists in words that would be devastating if they were applied to his own career and its sponsors: "The fact is that virtually every successful academic scientist is a ward of the federal government. One cannot do the research necessary to publish enough to be awarded tenure without appealing to one or another agency for considerable financial support.... Yet these and other agencies have their own political agendas."

A critical point that Michaels chooses to ignore is that all research sponsored by the federal government is subjected to the exacting requirements of scientific proof. In what is called the "refereed" literature, one's research peers systematically review an article as a condition of publication. By contrast, private, industry-funded research is not necessarily peer-reviewed and is frequently published in industry journals without undergoing this kind of rigorous scientific scrutiny.

But Michaels's argument founders on a far more obvious rock: If federally funded scientific research is merely a conspiracy to milk the public, why does the government permit it? And why, of all things, would it want to encourage reports of climate change? The federal government is already facing far more problems today than it has resources to handle. What conceivable reason would it have for funding scientists to exaggerate evidence of a coming climate catastrophe? Last time I looked, Congress had not approved any funding for agencies to discover remote, highly implausible new crises that would require greater public expenditures. Although Michaels has said he supports continued federal research funding, his testimony was cited by the chairman of the House Subcommittee on Energy and the Environment as a basis for cutting programs critical to monitoring climate change.

Questions of his funding aside, Michaels's statements have frequently blurred the roles of scientist and propagandist for his and his supporters' conservative political views. These views include bashing the United Nations.

In the spring 1996 issue of *World Climate Report*, Michaels reviewed the federal government's 1996 State of the Climate document. "If this is an official document," Michaels wrote, "there's no doubting that our federal government is a principal broadcast organ for the views of the United Nations IPCC.... It's obvious that the U.N. is viewed by the current administration as the defining entity for our climate."

Michaels's connections were further clarified in an article he authored in a 1993 issue of *World Climate Review*. This extensive article was essentially a retread of the Western Fuels video touting the beneficial effects of carbon dioxide. One source Michaels cited was Sherwood Idso's son, Keith E. Idso, a doctoral candidate at Arizona State University. Keith Idso is another skeptic who was hired by Western Fuels to testify at the St. Paul hearing.

Idso's testimony in St. Paul provided a moment of public embarrassment to his coal sponsors and a touch of comic relief for the audience. On the stand, he was asked about an article he had written titled "The Greening of the Planet." The article, which had appeared in a magazine called *The New American*, detailed in a fairly clinical scientific style his experiments on the effects of enhanced carbon dioxide on sour-orange trees. But it concluded with a startling burst of political rhetoric: "This good news [about enhanced carbon dioxide] is not what those intent on destroying our freedoms and imposing their will on the nations of the earth want us to hear, and they skillfully promote alternative voices to confuse the issue. The truth, however, will not be suppressed."

Assistant Attorney General Jonathan Wirtschafter asked Idso on the witness stand, "Mr. Idso, do you know if *The New American* is published by an advocacy group or a research institute?"

"I know it's not a scientific magazine," Idso replied. "It's something in the popular press."

"Is it published by an advocacy group of some sort?" Wirtschafter asked.

"I don't know if it's advocacy. I know it's some political type organization."

"What organization is that?"

"I can't remember," Idso said. "Some kind of society, I think."

"Was it the John Birch Society?" Wirtschafter asked.

Idso conceded that it was.

What is so extraordinary about the public career of Pat Michaels is that even after his initial association with the extremely cynical coal-funded campaign known as ICE, even after his publication of two journals financed by the coal industry, even after his receipt of money from such flagrantly ideological sources as the largest funder of the Wise Use movement and his use of source material published by the John Birch Society, he has nonetheless appeared as a star witness at several congressional hearings, most notably before the House Science Committee. There Michaels's testimony has been accorded more scientific credibility than that of scientists like Dr. Jerry Mahlman, director of NOAA's Geophysical Fluid Dynamics Laboratory at Princeton University; Dr. Michael MacCracken, a leading climate modeler at Lawrence Livermore National Laboratory for twenty-five years and later director of the largest federal climate science effort, the U.S. Global Change Research Program; and Dr. Robert Watson, cochair and lead author of the 1995 IPCC report on the impacts and uncertainties of global climate change.

The case of Robert Balling is equally intriguing. A geographer by training, much of Balling's research prior to 1990 focused on hydrology, precipitation, water runoff, and other southwestern water- and soil-related issues. Since 1991, however, the year he was solicited by Western Fuels, Balling has emerged as one of the most visible and prolific of the climate change skeptics.

Beginning with his work for the ICE campaign, Balling has also received, either alone or with colleagues, nearly $300,000 from coal and oil interests in research funding, which he has never voluntarily disclosed. In his collaborations with Sherwood Idso, Balling has received about $50,000 in research funding from Cyprus Minerals, as well as a separate grant of $4,900 from Kenneth Barr, who at the time was CEO of Cyprus. The German Coal Mining Association has provided about $80,000 in funding for Balling's work. The British Coal Corporation has kicked in another $75,000. Balling disclosed his industry funding under oath during the administrative hearings in Minnesota in 1995.

Given the obvious economic interests of OPEC in the climate debate, it is not surprising that Balling has also received a grant of $48,000 from the Kuwait Foundation for the Advancement of Science, as well as unspecified consulting fees from the Kuwait Institute for Scientific Research.

Balling's 1992 book, *The Heated Debate*, was published by a conservative think tank, the Pacific Research Institute, one of whose goals is the large-scale repeal of environmental regulations. Balling's book was subsequently translated into Arabic and distributed to the governments of the OPEC nations. The funding for this edition of his book was provided by the Kuwait Institute for Scientific Research.

Research funded by industry is not, to be sure, necessarily tainted. But public disclosure of industry funding is of critical importance, so that the research can be reviewed for possible bias. That disclosure requirement is mandatory in other areas of science. If a medical researcher's work is funded by, say, a pharmaceutical company, professional ethics demand that such funding be disclosed in a tagline, when the work is published in the *New England Journal of Medicine* or the *Journal of the American Medical Association*. It is unfortunate that the same standards of scientific and professional ethics do not extend to the field of climate science.

In late 1995 Balling authored an op-ed piece in *The Wall Street Journal* headlined "Keep Cool About Global Warming." Here he attacked the integrity of the IPCC, declaring that the panel's summaries are written "by a few group leaders, and it opens the door for slanting the underlying message of the comprehensive document. News accounts [based on those summaries] misrepresent reality when they use selective information, offer worst-case scenarios and make claims about increased confidence in the scientific community about predictions of potentially catastrophic climate changes."

It is understandable that a reader of *The Wall Street Journal*—say, a civic-minded executive—would be comforted to hear that concerns about global warming are overstated, especially given the tagline that identifies Balling as director of the Office of Climatology at Arizona State University. I doubt that the same reader would be quite as sanguine, however, if he knew that some of Balling's work was underwritten by German and British coal interests and by the government of Kuwait.

Among the skeptics, Dr. S. Fred Singer stands out for being consistently forthcoming about his funding by large oil interests. On a 1994 appearance on the television program *Nightline*, Singer did not deny having received funding from the Reverend Sun Myung Moon (to whose newspaper, *The Washington Times*, he is a regular contributor and whose organization has published three of his books). Nor has he apologized for his funding from

Exxon, Shell, ARCO, Unocal, and Sun Oil. Singer's defense is that his scientific position on global atmospheric issues predates that funding and has not changed because of it.

This interesting point raises an equally interesting question. What would happen if the climate skeptics just happened to stumble on a piece of evidence confirming that global warming is indeed intensifying? Would they be willing to alter the direction of their research at the risk of cutting off their industry funding? Such a situation, to say the least, would provide them with a very serious personal and professional conflict of interest. Fortunately for Singer, Michaels, and Balling, such a situation has never apparently arisen.

In early 1995, several months before one round of international climate negotiations, Singer proposed to an oil industry public relations outlet a $95,000 project in which he would mount a series of panels, lectures, and publications to "stem the tide towards ever more onerous controls on energy use." The project was intended to publicly counter the findings of the IPCC.

Indeed, the project bears more than a faint resemblance to the coal industry's ICE campaign. Singer's proposed oil-company-sponsored public education program would prepare a

> scientifically sound and persuasive critique of the IPCC summary.... Next we would distribute the Critique widely...and publish a Statement of Support by a hundred or more climate experts. This Statement could then be quoted or reprinted in newspapers. Our proposal envisages assembling a panel of about five distinguished scientists/technologists. This panel would issue a Release pointing up the IPCC Critique and conduct press briefings to defend its conclusions. If funding can

be provided without delay, the panel...could issue its Release...during or before the [round of international climate negotiations] meeting in New York.

In his wind-up, Singer warned the oil companies that they face the same threat as the chemical firms that produced chlorofluorocarbons (CFCs), a class of chemicals that were found to be depleting the earth's protective ozone layer. "It took only five years to go from...mandating a simple freeze of production [of CFCs] at 1985 levels, to the 1992 decision of a complete production phase-out—all on the basis of quite insubstantial science," Singer wrote.

Contrary to his assertion, however, virtually all relevant researchers say the link between CFCs and ozone depletion rests on unassailable scientific evidence. Three months later, as if to underscore the CFC-ozone connection, the research director of the European Union Commission announced that the previous winter's ozone loss would result in about 80,000 additional cases of skin cancer in Western Europe. Shortly thereafter, the three scientists who discovered the CFC-ozone link were awarded the Nobel Prize for chemistry. But that did not faze Singer, who proceeded to attack the Nobel committee in the pages of Reverend Moon's *Washington Times*. In his November 1995 article, headlined "Ozone Politics with a Nobel Imprimatur," Singer declared that "The Swedish Academy of Sciences has chosen to make a political statement.... The selection committee evidently decided to reward global environmentalism rather than a fundamental advance in the basic science of chemistry."

The following month, researchers for the World Meteorological Organization announced that the ozone hole over Antarctica had grown at an unprecedented rate in 1995, covering an area twice the size of Europe at its peak in October. A one-percent-per-day decline in the ozone layer during August had caused the ozone hole to expand more rapidly than in any previous year, reaching a maximum of 7.7 million square miles, the researchers reported.

Singer's tantrum against the Nobel committee would be laughable—except that his views exert serious influence, especially on conservative politicians. Based in part on Singer's work, House majority whip Tom DeLay and Representative John Doolittle are making an effort to withdraw U.S. participation in the Montreal Protocol—the international compact that mandates an end to production of the chemicals that destroy the ozone layer. Despite the remarkable international consensus on the Montreal Protocol, DeLay used Singer's pronouncements to attribute it merely to "a media scare."

Not long after launching his attack on the Swedish Academy, Singer revisited the climate debate in another article in *The Washington Times*, again using the vocabulary of propaganda. "Early this year," he wrote,

> *The New York Times* ominously warned ''95 Hottest Year on Record,' implying the existence of a strong global warming trend, presumably caused by the greenhouse effect of increasing atmospheric carbon dioxide. The story did not reveal that the headline was based on an earlier British Meteorological Office release that used data from only the first 11 months of 1995. December turned out to be cold, making 1995 an average year. The scary headlines, echoed by a *Newsweek* article, created only a minor stir—appearing just as the January blizzards began to hit. Nice try, fellows: bad timing!

Unfortunately for him, Singer made two mistakes. First, severe winter weather is perfectly consistent with global warming. One effect of climate change is to produce more extreme local temperatures—leading to hotter hots, colder colds, and more severe snowstorms. Global warming has not yet eradicated the seasons of the year, even if it may be affecting their timing. Although this point is part of our basic knowledge of climate change, it appears to have eluded Singer.

Singer was also wrong about the weather record. According to the World Meteorological Organization, even when the unusually cold December temperatures are included, 1995 turns out to have been the hottest year in recorded weather history. The World Meteorological Organization also notes that there were more hurricanes over the Atlantic than in any year since 1933 and that the Antarctic ozone hole lasted longer than in any previous year.

Despite his obvious misstatements of fact, however, Singer continues to be widely quoted by the news media.

I watched Al Gore deliver this speech on the floor of the Kyoto conference in December of 1997. It was notable as much for the scene as for the content. The conference had reached an impasse, with the Europeans wanting to go further in regulating carbon and the American delegation balking. But when Gore uttered the diplomatic phrase "I am instructing our delegation right now to show increased negotiating flexibility," the hall erupted in applause. And indeed a Kyoto accord was scraped together at the last minute, as officials of the convention center were literally shooing delegates out the door to make room for a trade show. That may have been the high-water mark of international progress on climate change, however, for the U.S. Congress never ratified the agreement, making Kyoto a poignant milestone on the political road to nowhere.

# Remarks at the Kyoto Climate Change Conference

Al Gore
*1997*

Prime Minister Hashimoto and President Figueres, President Kinza Clodumar, other distinguished heads of state, distinguished delegates, ladies and gentlemen.

It is an honor to be here at this historic gathering, in this ancient capital of such beauty and grace. On behalf of President Clinton and the American people, and our U.S. negotiator, Ambassador Stu Eizenstat, I salute our Japanese hosts for their gracious hospitality, and offer a special thank you to Prime Minister Hashimoto, and to our chairs, Minister Ohki, and Ambassador Estrada, for their hard work and leadership.

Since we gathered at the Rio Conference in 1992, both scientific consensus and political will have come a long way. If we pause for a moment and look around us, we can see how extraordinary this gathering really is.

We have reached a fundamentally new stage in the development of human civilization, in which it is necessary to take responsibility for a recent but profound alteration in the relationship between our species and our planet. Because of our new technological power and our growing numbers, we now must pay careful attention to the consequences of what we are doing to the Earth—especially to the atmosphere.

There are other parts of the Earth's ecological system that are also threatened by the increasingly harsh impact of thoughtless behavior:

The poisoning of too many places where people—especially poor people—live, and the deaths of too many children—especially poor children—from polluted water and dirty air; the dangerous and unsustainable depletion of ocean fisheries; and the rapid destruction of critical habitats—rainforests, temperate forests, boreal forests, wetlands, coral reefs, and other precious wellsprings of genetic variety upon which the future of humankind depends.

But the most vulnerable part of the Earth's environment is the very thin layer of air clinging near to the surface of the planet, that we are now so carelessly filling with gaseous wastes that we are actually altering the relationship between the Earth and the Sun—by trapping more solar radiation under this growing blanket of pollution that envelops the entire world.

The extra heat which cannot escape is beginning to change the global patterns of climate to which we are accustomed, and to which we have adapted over the last ten thousand years.

Last week we learned from scientists that this year, 1997, with only three weeks remaining, will be the hottest year since records have been kept. Indeed, nine of the ten hottest years since the measurements began have come in the last ten years. The trend is clear. The human consequences—and the economic costs—of failing to act are unthinkable. More record floods and droughts. Diseases and pests spreading to new areas. Crop failures and famines. Melting glaciers, stronger storms, and rising seas.

Our fundamental challenge now is to find out whether and how we can change the behaviors that are causing the problem.

To do so requires humility, because the spiritual roots of our crisis are pridefulness and a failure to understand and respect our connections to God's Earth and to each other.

Each of the 160 nations here has brought unique perspectives to the table, but we all understand that our work in Kyoto is only a beginning. None of the proposals being debated here will solve the problem completely by itself. But if we get off to the right start here, we can quickly build momentum as we learn together how to meet this challenge. Our first step should be to set realistic and achievable, binding emissions limits, which will create new markets for new technologies and new ideas that will, in turn, expand the boundaries of the possible and create new hope. Other steps will then follow. And then, ultimately, we will achieve a safe overall concentration level for greenhouse gases in the Earth's atmosphere.

This is the step-by-step approach we took in Montreal ten years ago to address the problem of ozone depletion. And it is working.

This time, success will require first and foremost that we heal the divisions among us.

The first and most important task for developed countries is to hear the immediate needs of the developing world. And let me say, the United States has listened and we have learned.

We understand that your first priority is to lift your citizens from the poverty so many endure and build strong economies that will assure a better future. This is your right: it will not be denied.

And let me be clear in our answer to you: we do not want to founder on a false divide. Reducing poverty and protecting the Earth's environment are both critical components of truly sustainable development. We want to forge a lasting partnership to achieve a better future. One key is mobilizing new investment in your countries to ensure that you have higher standards of living, with modern, clean and efficient technologies.

That is what our proposals for emissions trading and joint implementation strive to do.

To our partners in the developed world, let me say we have listened and learned from you as well. We understand that while we share a common goal, each of us faces unique challenges.

You have shown leadership here, and for that we are grateful. We came to Kyoto to find new ways to bridge our differences. In doing so, however, we must not waiver in our resolve. For our part, the United States remains firmly committed to a strong, binding target that will reduce our own emissions by nearly thirty percent from what they would otherwise be—a commitment as strong, or stronger, than any we have heard here from any country. The imperative here is to do what we promise, rather than to promise what we cannot do.

All of us, of course, must reject the advice of those who ask us to believe there really is no problem at all. We know their arguments; we have heard others like them throughout history. For example, in my country, we remember the tobacco company spokesmen who insisted for so long that smoking did no harm. To those who seek to obfuscate and obstruct, we say: we will not allow you to put narrow special interests above the interests of all humankind.

So what does the United States propose that we do?

The first measure of any proposal must be its environmental merit, and ours is environmentally solid and sound.

It is strong and comprehensive, covering all six significant greenhouse gases. It recognizes the link between the air and the land, including both sources and sinks. It provides the tools to ensure that targets can be met—offering emissions trading, joint implementation and research as powerful engines of technology development and transfer. It further reduces emissions—below 1990 levels—in the years 2012 and beyond. It provides the means to ensure that all nations can join us on their own terms in meeting this common challenge.

It is also economically sound. And, with strict monitoring and accountability, it ensures that we will keep our bond with one another.

Whether or not agreement is reached here, we will take concrete steps to help meet this challenge. President Clinton and I understand that our first obligation is to address this issue at home. I commit to you today that the United States is prepared to act—and will act.

For my part, I have come here to Kyoto because I am both determined and optimistic that we can succeed. I believe that by our coming together in Kyoto we have already achieved a major victory, one of both of substance and of spirit. I have no doubt that the process we have started here inevitably will lead to a solution in the days or years ahead.

Some of you here have, perhaps, heard from your home capitals that President Clinton and I have been burning up the phone lines, consulting and sharing new ideas. Today let me add this. After talking with our negotiators this morning and after speaking on the telephone from here a short time ago with President Clinton, I am instructing our delegation right now to show increased negotiating flexibility if a comprehensive plan can be put in place, one with realistic targets and timetables, market mechanisms, and the meaningful participation of key developing countries.

Earlier this century, the Scottish mountain climber W.H. Murray wrote: "Until one is committed there is hesitancy, the chance to draw back, always ineffectiveness. Concerning all acts of initiative...there is one elementary truth, the ignorance of which kills countless ideas and splendid plans: that the moment one definitely commits oneself, providence moves, too."

So let us press forward. Let us resolve to conduct ourselves in such a way that our children's children will read about the "Spirit of Kyoto," and remember well the place and the time where humankind first chose to embark together on a long-term sustainable relationship between our civilization and the Earth's environment.

In that spirit, let us transcend our differences and commit to secure our common destiny: a planet whole and healthy, whose nations are at peace, prosperous and free; and whose people everywhere are able to reach for their God-given potential.

Thank you.

*The difficulties of making real progress against climate change became ever clearer in the years after Kyoto. For one thing, an ascendant right wing in the United States refused to act on the issue. At the same time, the explosive emergence of China as a world economic power made the calculus of carbon control much harder. By the early part of the millennium, the Chinese were famously opening a new power plant every week, most of them powered by coal. Though their emissions still lagged far behind America's in per capita terms, they were increasing so fast that the Chinese were suddenly key players in any negotiations. To understand why Beijing was committed to increasing energy use so quickly, you needed to know something about the depth of rural poverty in China. Mark Hertsgaard, a veteran environmental journalist, provided some unforgettable snapshots in his important book* Earth Odyssey.

# Is Your Stomach Too Full?

from *Earth Odyssey*

Mark Hertsgaard
*1998*

> "Master," I cried, "who are these people
> By black air oppressed?"
> —Dante, *The Inferno*

Of all the things you need when traveling around the world, luck may be the most decisive, and meeting Zhenbing was certainly lucky. I must confess, it didn't feel like luck when my original interpreter in China abruptly abandoned the project thirty-six hours after I reached Beijing. But then Zhenbing appeared, introduced by a mutual friend, and I soon realized fate had done me a favor. A thirty-year-old economics instructor at one of Beijing's major universities, Zhenbing had a ready laugh, high cheekbones in a long, angular face, and glasses that did not quite hide the slightly off-center gaze of his right eye. He had a few weeks free before next semester's classes began, and the idea of traveling around within his native land appealed to his adventurous spirit. His English was strong (he had recently returned from two years of graduate study at one of Europe's most prestigious academic institutions), and he was an experienced interpreter. Nor was he troubled by my plan to operate below the government's radar screen, traveling as a tourist to evade the media minders who monitored all permanently stationed foreign reporters in China.

Zhenbing's greatest strength as an interpreter was that he could talk to anybody, from lowly peasants to high party leaders.

He was just one of those people everyone likes on the spot. Part of his secret was that he was extremely funny and laughed a lot, especially at himself, a talent he had plenty of opportunities to display, for he had without question the worst sense of direction of anyone I met in all my travels. This is generally not a talent one looks for in interpreters, and it wasn't long before I took over all our navigating. But Zhenbing was a good sport, and to amuse ourselves at the end of the day, we would sometimes compare ideas for how to get back to the hotel or train station. At these moments Zhenbing would look around intently, furrow his brow in concentration, and finally point decisively in one direction or another. But his eyebrows, raised in bravely hopeful fashion, betrayed his inner uncertainty, and when he learned he had chosen wrongly once again, he would burst into uproarious laughter.

Above all, Zhenbing was a great, sometimes endless, talker and an enthusiastic storyteller. I can't count the number of times he struck up a conversation with the person next to him in a bus station or train compartment, and soon it wasn't just the two of them but a whole crowd of previously silent strangers chattering back and forth like long-lost relatives. Zhenbing was invariably at the center of things, with the crowd hanging on his every word—laughing, interjecting questions and commentary, and generally having a fine old time. It was a bit like watching the Pied Piper in action, except that Zhenbing's charm captivated grownups and children alike—and gave me access to countless candid conversations that I never would have heard otherwise.

Another advantage of traveling with Zhenbing was that he saw life the way a peasant did, which is to say, the way the vast majority of Chinese did. Historically, ninety percent of China's population had lived in the countryside, and even after the great urban migrations sparked by the economic reforms of the

1980s, the figure remained over seventy percent. Born in 1966, Zhenbing had grown up very poor, passing his first twenty years in a small village northwest of Beijing near the border with Inner Mongolia. The second of three brothers, he was fourteen years old before he got his first pair of shoes. His family inhabited a mud straw hut and was too poor to buy coal in winter; for heat, they burned straw. They did this in a climate as cold as Boston's or Berlin's, with winter temperatures that dipped thirty degrees below zero. "Often the straw was not enough, so the inside wall of the hut became white with icy waterdrops, like frozen snow," Zhenbing recalled. "In my village, when a girl was preparing to marry, the first thing the parents checked was, will the back wall of the would-be son-in-law be white or not? If not white, they approved the marriage, because that meant his family was wealthy enough to keep the house warm."

Zhenbing was an invaluable reality check for an outsider like myself investigating China's environmental crisis. He never made an issue of his background, perhaps because it was such a common one in China. But in ways large and small he helped me understand that, while there were plenty of things the Chinese masses might not like about their existence, by far their biggest complaint was being miserably poor, and they would put up with a great deal of aesthetic or environmental unpleasantness to escape that poverty.

As recently as 1950, the average life expectancy in China was thirty-nine years, a level not seen in Europe since the Industrial Revolution. And many Chinese of the 1990s still had firsthand memories of suffering through the greatest man-made disaster of the twentieth century, the famine caused by Mao Zedong's Great Leap Forward campaign. As Jasper Becker documented in his powerful book, *Hungry Ghosts*, Mao's famine killed some thirty million people between 1959 and 1961—more than Hitler's and Stalin's death tolls combined—and brought misery, starvation, and even cannibalism to virtually all parts of rural China.

Chinese life spans averaged about seventy years by the mid-1990s, yet hundreds of millions still lived in desperate poverty. In one village Zhenbing and I visited in Sichuan province, on an early January day when my feet were barely comfortable inside polar-insulated hiking boots, I watched a grim-faced peasant woman seated on a rocky river bank, her bare feet dangling in the frigid water as she washed her family's clothes. On the other side of the village, three young children amused themselves with the only toy they had, a plastic water bottle filled with pebbles, which they pulled around like a wagon on a string. Behind them, a barefooted man stamped around on a pile of loose, moist coal, looking like an eighteenth-century European peasant crushing grapes for wine. In fact, he was manufacturing—by foot, as it were—the briquettes of fuel whose carcinogenic combustion would provide what little heat he and his neighbors enjoyed in their windowless mud huts.

Back in Beijing, Zhenbing agreed after some coaxing one day to show me his dormitory room at the university. It was on the third floor of a long, barnlike building. Out front, scores of bicycles slouched against one another like tired children. From the moment we pushed through the scarred wooden doors on the building's ground floor, we were surrounded by the smell of toilets, stale air, and general uncleanliness. The concrete stairs were filthy, the walls splattered with all manner of stains and dirt—nothing had been painted for decades. Though it was two o'clock in the afternoon, it was very dark inside the dorm. Following Zhenbing up to the second floor, I saw that the only light came from dim, naked bulbs dangling from the ceiling at intervals of fifty feet or so. The dusty, shadowy corridors were like obstacle courses, overflowing with boxes, cooking gear, desks, and whatever other belongings could not fit inside the rooms.

The odor was rather worse upstairs, because the toilets led directly off the corridor, and their doors were wide open. Along one wall of the men's room was a stand-up urinal that did not seem to flush. Along the opposite wall were six squat toilets that were open in front—"There is never privacy here," Zhenbing complained—and separated only by concrete partitions. There was no toilet paper, only a hole in the concrete floor and a small red lever with which to flush away waste. One recent visitor had neglected that task, leaving a deposit of frighteningly large proportion to fester and stink. There were no bath or shower facilities; all washing was done in a room down the hall, where two rows of concrete tubs offered cold water only. Mirrors, counter space, towel racks, and similar luxuries were nonexistent.

Inside Zhenbing's room the smell was also unpleasant, for which he quickly apologized. The walls were grimy and flaking from mildew. When I tried to open the one tiny window, I found it had been broken and the missing piece literally papered over with a sheet of notebook paper. The room was six paces wide and four paces long, yet it somehow contained two cast-iron bed frames, a metal bookcase, two wooden desks, a coat stand, and a set of rickety shelves. The light was again a single overhead bulb, and the beds were as hard as stone: a wooden plank topped by a bamboo mat and thin cotton quilt.

The place felt shabby and grim and spirit-crushing, but Zhenbing pointed out that it was in fact a relative privilege. As a junior faculty member, he had this room to himself. Ph.D. students would live three together in the same space, while master's students were packed four together and undergraduates six to eight. When I asked whether these humble conditions ranked above those of the rural majority in China, Zhenbing replied, with some bitterness, "Oh, yes, definitely. You must remember, the government is afraid of students getting upset, so they treat us relatively well. We have access to electricity, to running water, to central heating. Paradise!"

Zhenbing was riled up now, and as we headed off campus for our next appointment his frustration burst forth in a tirade that alternately attacked and defended the government. "This is why economic development is the most important goal for China," he said. "It is more important than environment, or human rights, or the other issues the Western media and governments complain about. You may think it is propaganda, but most Chinese support the government for building the Three Gorges dam [a controversial project under construction on the Yangtze River]. That is no business of yours! You may complain about our contributions to global warming or the ozone hole, because those issues affect everyone on earth. But how much pollution we make, how many trees we cut or dams we build is nobody's business but ours."

Among his many other talents, Zhenbing also possessed world-class sleeping abilities. The man could fall asleep anywhere, anytime, for short periods or long, then wake up alert, cheerful, and eager to chat. One day, on a train from Shenyang to Beijing, he slept for more than eight hours seated directly beneath a loud-speaker that blared tinny Chinese pop music, inside a compartment where passengers were crammed knee-to-knee and shoulder-to-shoulder, while the aisle overflowed with people pushing and shoving toward the rear to refill tea thermoses or visit the reeking toilets. Zhenbing told me he learned his sleeping skills in college, by necessity, while sharing a room with seven other students. (Endless college card games also made Zhenbing a good enough gambler that years later, in the West, he supplemented his income by visiting casinos and regularly beating the house, but that's another story.)

My most vivid memory of our train ride to Beijing, though, was of all the spitting that went on. I was wedged between a sniffling peasant girl on one side and her older brother or cousin on the other. Zhenbing and I had "hard-seat" tickets—the lowest class, and the only ones available. I was a curiosity to these peasants, and to show friendship the young man offered me a few of the sunflower seeds he and his family of seven were munching. Upon finishing his own seeds, he washed them down with a swig of tea and then, with a deep hocking sound, summoned from his throat a prodigious gob of phlegm, which he casually spat onto the floor in front of us. Out of the corner of my eye, I watched as he then reached out his foot and rubbed the spit into the floor, as if stamping out a cigarette. It was 8:15 in the morning, there were fourteen more hours to Beijing, and there was lots more spittle loosed throughout that packed compartment before we got there.

Everyone seemed to spit in China—on the sidewalk, in the classroom, on the train, in restaurants, wherever. Middle-aged housewives, rowdy teenagers, toothless old men, beautiful young women—the habit was universal. The communists had tried to eradicate spitting when they came to power in 1949; it was one of their first exhortations to the masses. They failed. Spitting lived on in the 1990s because it was a habit of peasant life, and the vast majority of Chinese were still peasants or only one generation removed. The habit itself apparently derived from the basic conditions of peasant life, which included rampant lung infections and other respiratory diseases. These, in turn, resulted from a historical fact with enormous environmental implications:

for centuries, Chinese peasants had lived with very little heat in wintertime, so they were frequently sick. They burned wood—if they were lucky—or straw and cropstalks, as Zhenbing's family did. In the 1990s, Chinese peasants still relied on such "biomass" fuels for seventy percent of their energy consumption (and found themselves short of fuel between three and six months out of the year).

Coal therefore represented a great advance for the Chinese people; it kept a body much warmer. But it did so at terrible cost. Coal smoke was the most important element in the air pollution that was killing at least 1.9 million people a year in China, according to the World Bank. "Pollution is one reason chronic obstructive pulmonary disease—emphysema and chronic bronchitis—has become the leading cause of death in China, with a mortality rate five times that in the United States," reported the bank. Outdoor air pollution was second only to cigarette smoking as a cause of lung cancer in China's cities, where lung cancers had increased 18.5 percent since 1988. Coal smoke was also the main component of the indoor pollution from home stoves that caused most rural lung cancers, especially among women.

Like the automobile, nuclear fission, and so many other technologies in human history, coal burning in China was both a blessing and a curse. Coal had brought China into the industrial era and enabled the Chinese masses to be warm in winter for the first time in history. At the same time, it had poisoned the air and water, not to mention people's lungs, beyond description. The huge amounts of coal burned in China (especially in the north, which relied on coal for winter heat), along with the primitive technologies often employed, gave large northern cities such as Beijing and Shenyang some of the filthiest air on earth. The levels of total suspended particulates (TSP) routinely climbed as high as 400 and 500, even 800 in winter—four to nine times greater than World Health Organization guidelines.

I had breathed plenty of bad air in my travels, but none like Beijing's; it made Bangkok's air seem merely unpleasant. According to He Kebin, vice chairman of the Department of Environmental Engineering at Tsinghua University, approximately seventy-five percent of Beijing's air pollution in winter was caused by burning coal, the fuel that heated and powered the city. I had been warned in advance about Beijing's air. One journalist told me about a Western diplomat friend, an avid jogger who refused to give up his habit during a two-year stint in the capital. When the diplomat finally returned to his home country and underwent a routine medical checkup, his physician told him he absolutely had to quit smoking. Need I add that cigarettes had never passed this man's lips?

Nevertheless, when I arrived in Beijing, I had to wonder if all that talk had been mere journalistic exaggeration. My first morning in town, I bounced out of bed and eagerly headed outside for my first walk in the People's Republic. The temperature that December day was a bracing nineteen degrees, but the real surprise was the brilliantly blue and sunny sky. There were no signs of pollution anywhere. How could this be?

It was just before eight o'clock, and the four-lane boulevard outside my hotel was crowded with commuters, a stream of humanity so dense and fast-moving that at first I could only stand back and watch. As wheezing buses rumbled to a halt at the side of the boulevard, bunches of pedestrians would dash shouting and laughing into the throng already waiting at the bus stop in hopes of squeezing their way aboard the crushingly packed vehicle. A few people traveled by taxi or car, but the vast majority was on bicycles, usually the stolid black Chinese model called

Flying Pigeons. There were also lots of three-wheeled cargo bikes. They had long, drooping gear chains hanging inches off the ground and, in back, wooden flatbeds that carried everything from bulging sacks of fruit and vegetables to freshly skinned sides of pork, to couches, toilets, televisions, and small mountains of crushed cardboard boxes destined for recycling. I was especially intrigued by bikes carrying the coal briquettes locals called "honeycombs" (because of the holes drilled in the briquettes to encourage cleaner burning). Round, black, the size of small coffee cakes, the honeycombs were stacked by the hundreds into squat pyramids and sold off the carts for burning in the home stoves of the poor. Honeycombs were said to be the cause of much of China's air pollution, but where was that pollution?

It took me forty-five minutes to walk a square city block that morning, partly because the wind was so fierce that I sometimes had to duck into doorways to escape it. One sidewalk was occupied by dozens of street vendors, each swathed in countless layers of clothing and hunched behind frozen piles of nuts, grains, eggs, fruits, or shriveled white cabbages. Whatever romantic expectations I had had of Beijing's architecture were quickly extinguished by the pervasive drabness of the place. The buildings were square, ugly concrete boxes, and Beijing, like most of China, was the most litter-strewn place I had ever visited. Everywhere I looked there was trash—plastic bags, peanut shells, cigarette boxes, food cartons, construction site refuse—and the gusting winds sent it blasting along the streets like pellets from a shotgun.

I returned to my hotel chilled, exhilarated, and bewildered, and I spent the rest of the day plagued by a fear familiar to all journalists: had I come in search of a nonexistent story? Where was all the air pollution?

That afternoon, a government press aide proudly explained that the government had moved all the city's heavy industry out of the downtown area. It sounded plausible, and I subsequently learned that some factories had indeed been relocated. But it turned out that Beijing was, in fact, rarely graced by blue skies in winter, except immediately after Siberian winds had roared through and flushed away all the smog. By chance, such winds had struck the night I arrived and continued blowing throughout the following day. As a taxi driver told me that second evening, the only reason Beijing's air did not look dirtier was that "it's very windy today. If there were no wind, you'd notice [the pollution] very strongly."

Sure enough, the winds calmed the next day, a Friday, and over the following ten days I witnessed the sickening descent of Beijing into a city of murk and gloom.

At noon on Saturday, after just twelve hours of still air, I took a bus across town to a luncheon interview, traveling the main west–east boulevard through Tiananmen Square. Straight above me, the sky was still blue, but in the distance a fuzzy, pale gray layer of smog already frosted the skyline. When I came back outside four hours later, the layer had nearly doubled in thickness, its blurry density giving the sky an otherworldly aspect as it melted into a sunset of vivid pinks and yellows. The pollution accumulated with each passing day, and by Thursday I was used to waking up to a dull, gray-white haze that rested on the city skyline like a lid on a wok.

The haze would grow palpably worse through the course of a day, as countless thousands of boilers were fired up and internal combustion engines spewed exhaust. On Thursday, for example, Zhenbing and I were riding south on the second ring road that skirts Purple Bamboo Park on its way around the western edge of town. It was about 4:30 in the afternoon, and we were at the head of a long, flat stretch of highway. Our eyes should have been drawn to the Chinese national television (CCTV) tower, by far the tallest structure in the city, which lay directly in front of us, about four miles away. But by this hour the smog had become so thick that what had been a basically sunny day at noon now looked overcast and dark. Only because I knew the CCTV tower had to be up ahead could I faintly make out its needle-nosed outline against the sky. To my right, striding down the sidewalk, I saw a tall, young woman in a smart black overcoat. Behind her, two young girls wearing bright red and yellow athletic suits pedaled bicycles. Farther back, a wizened old fellow in a blue Mao cap was bent over an upturned bike, trying to repair it. I looked back toward the CCTV tower and realized that to anyone at the tower, the air around here must look just as bad as the air up there appeared to me, and I swallowed hard at how much poison was entering our lungs.

On Friday morning, I took a taxi to the National People's Congress on Tiananmen Square. Passing through the larger intersections of Beijing, I looked both ways down the cross streets but could see no farther than half a block; beyond that, an impenetrable gray mass concealed everything. When I reached Tiananmen Square, at 8:45, the sun hung white and barely visible above the southern gate to the Imperial Palace, like a weak lightbulb in a barroom full of cigarette smoke. Gazing north, past Mao's mausoleum and the site of the 1989 massacre, I could not see the far end of the square, much less the Forbidden City beyond it. The pedestrians crossing the square were like spectral figures—half ghost, half flesh—as they disappeared into the gritty mist.

It had now been a week since the Siberian winds had cleansed Beijing, and I craved a respite from the increasingly filthy air.

On Sunday I took a public bus to the Great Wall of Badaling, seventy-five kilometers north of the capital. Though genuinely foggy at the wall, the fresh air felt wonderful descending into my lungs, and northern winds even brought a cheering patch of blue sky by midafternoon. But not to Beijing. Back in the city, I stepped off the bus into a grimy dusk. Beneath my feet, whorls of coal dust spiraled across the sidewalk like black snow flurries.

How did people stand it? The bicycle delivery drivers who strained to pedal their overloaded carts uphill amid air so polluted it nearly glowed—how did they do it? Zhenbing brushed aside such questions. "We are used to it," he said. "I have lived here for years, so my body has gotten used to this air."

I heard the same from countless other locals. Foul air was simply a fact of life in Beijing, they believed, an inevitable result of the sharp increase in motor vehicles, office buildings, neon lights, private shops, karaoke restaurants, and other forms of economic activity that had breathed new life into the city in recent years. Zhenbing did not enjoy breathing such air—he wasn't stupid. But to someone who had learned in childhood how trying life without coal could be, he saw pollution as the lesser of two evils. Observing Zhenbing's stoicism, I was reminded yet again of the young Ugandan at the park above the Nile, who defended damming the river because it meant electricity for his people. Zhenbing regarded extra pollution as a trade-off he was ready, even eager, to make in return for warmer apartments, busier factories, fewer power shortages, and a higher standard of living. Who could blame him?

Despite its terrible health effects, China's dependence on coal is bound to continue, if only because coal is one of the few natural resources China has in any abundance. No other country produces or consumes as much coal as China does. When I visited Datong, an ugly, low-slung city in Shanxi province known as China's coal mining capital, residents proudly told me time and again, "Datong sits on a sea of coal!" The Number 9 Coal Mine, ten miles from downtown Datong, produced ten thousand tons of coal a day, and there were fifteen mines of equal capacity in the near vicinity, plus hundreds of smaller privately owned ones. All day long, the roads around Datong were filled with sky-blue cargo trucks brimming with chunks of coal. I saw coal transported by virtually every other means imaginable, too: trains, smaller trucks, and wooden carts pulled by three-wheeled motorcycles and even by donkeys. The roadsides were crowded with members of the poorer classes, who used shovels, brooms, and bare hands to scoop fallen ore into baskets they balanced across their shoulders for the walk home. Although Datong's streets were almost literally paved with coal, I did not appreciate the sheer volume involved until the morning Zhenbing and I left town. Rolling south through Shanxi province, our train passed a number of huge, looming mountains of coal. The first mountain was easily seven stories tall and extended along the tracks for nearly two miles. On top of the mountain I saw a yellow bulldozer, shoving around the black ore and looking as tiny and inconsequential as a child's toy on a sandpile.

Coal accounts for three-quarters of China's total energy consumption, which is the other reason China's coal dependence is likely to last for decades: virtually the entire national infrastructure runs on coal. The exception is the transportation system, which is based on petroleum and human muscle. But three-quarters of electricity production and virtually all the factories and heating are coal-powered. To replace or upgrade this infrastructure—

the hundreds of electric power stations, the hundreds of thousands of boilers in factories and apartment buildings, the millions of honeycomb stoves—is an essential but necessarily long-term project, especially because Chinese of all classes are impatient to have economic progress today.

After all, this is a country where the overwhelming majority of people does not have a refrigerator, electricity shortages are constant, and one hundred million peasants live without any electricity whatsoever. "Electricity shortages exist in all the big cities, except Beijing," Lang Siwei, director of the Air Conditioning Institute of the Chinese Academy of Building Research, told me. "Shortages are very severe among Yangtze River cities, especially small cities, which often have no electricity during the daytime hours of peak demand. The only reason Beijing is spared such shortages is that electricity gets diverted from other cities to keep the capital running."

"In ninety percent of the villages there is electricity, but it can be cut off at anytime because of the pervasive shortages in the system," said Zhou Dadi, deputy director general of the State Planning Commission's Energy Research Institute. "Most rural families have electric light, and a few might also have a television and a small refrigerator. But not more than that, because the weak fuses and supply lines can't handle it."

One reason for the shortages, of course, is that demand for refrigerators and other consumer appliances has skyrocketed in recent years. Only seven percent of urban households had a refrigerator in 1985; by 1994, sixty-two percent did. During the same period, possession of color televisions increased from seventeen to eighty-six percent of households: clothes washers, from forty-eight to eighty-seven percent; and air conditioners, from zero to five percent. In the countryside, where more people lived but there had been much less development, the steepest increases were of electric fans and black-and-white televisions, which increased from approximately ten percent of households in 1985 to eighty-one and sixty-two percent, respectively, in 1994. Clothes washers, color TVs, and especially refrigerators remained rare in rural areas—approximately fifteen percent of households had one or more of these items—but here too the trend lines were climbing sharply.

Electricity is the fastest-growing part of China's energy demand. Its growth averaged nine percent a year from 1984 to 1994 and was projected to continue at no less than seven percent a year through 2000 and beyond. That means that China's total electrical generating capacity has to double every decade to keep pace. Thus, declared Lang Siwei, "electricity shortages are certain in China for the next five years. Whether they continue after that will depend on two things: whether we can improve our energy efficiency sufficiently, and whether we can build all the new power plants we have planned." According to Pan Baozheng, a senior engineer with the State Science and Technology Commission, China was planning to build thirty to sixty new electric power plants every year for the foreseeable future. The plants would have capacities of 300 to 600 megawatts each, and seventy-five percent of them would be coal-fired. China is "the biggest market for electric power plants in the world," Pan told me with pride.

All this economic growth has led experts to predict that China's total coal production, which reached 1.3 billion metric tons in 1996, would double if not triple by 2020 at the latest. Much will depend on how well China implements the energy efficiency and other infrastructure reforms mentioned earlier, according to

William Chandler, director of the Advanced International Studies Unit at the Battelle, Pacific Northwest Laboratories of the U.S. Department of Energy. "Our scenarios predict either a doubling of coal consumption with efficiency reforms, or a tripling without them," said Chandler, who has been regularly visiting China and collaborating with its energy planners for over a decade. These increases would occur "by 2015 if one assumes a continuation of rapid economic-growth in China, by 2020 with moderate growth," Chandler added.

Besides adding to China's problems with filthy skies and battered lungs, the additional coal burning would threaten the rest of the world by greatly increasing China's production of acid rain and greenhouse gases. In geographic contrast to the TSP [total suspended particulate matter] problem, acid rain is most pronounced in the southern half of China, because the sulfur content of southern coal is higher and southern soils are naturally more acidic. As with TSP, sulfur dioxide levels in much of China are many times greater than the World Health Organization guideline of forty to sixty micrograms per cubic meter annually. Among China's most afflicted cities were the coal center of Taiyuan, where the average level in 1996 was 230. Worst of all was Chongqing. Its 1996 average reading was 320, but in winter it climbed above 600!

Acid rain affects twenty-nine percent of the land area in China and causes $5 billion worth of damage every year. Emissions of sulfur dioxide—the source of acid rain—corrode buildings, encourage respiratory disease, and damage forests, lakes, and agriculture. In the Chongqing area, twenty-four percent of the 1993 vegetable crop was damaged by acid rain, with cereal crops suffering equivalent losses. In a country that is straining to feed itself, and where the rural majority is so poor it relies on wood and other biomass for seventy percent of its fuel, such losses are especially costly.

Acid rain is "quite a serious problem" for China, said Wang Wenxing, a senior adviser at the Chinese Research Academy of Environmental Sciences, "but in the future it will be much worse," because its effects can take decades to manifest. "In the United States," added Wang, "you can see the effects of your earlier industrialization today. In China, the industrialization is under way now, so the effects will be growing later." Since sulfur dioxide emissions can travel many hundreds of miles before falling to earth as rain, China's coal burning has dramatically worsened the acid rain problem in both Japan and South Korea; China is responsible for half of Japan's acid rain and eighty percent of South Korea's. In Japan, as in Europe and the United States, electric utility companies were forced in the 1970s and 1980s to add "scrubbers" that removed sulfur dioxide from their coal plants' exhaust. Tokyo is now funding an extensive program of technology transfer aimed at outfitting Chinese plants with their own scrubbers.

No such technical fix is possible with the carbon dioxide emissions that enhance the greenhouse effect, however. Coal is the most potent carbon dioxide producer of all fossil fuels: it emits twenty-five percent more carbon dioxide per unit of energy use than petroleum does, and nearly twice as much as natural gas. Although China relies on petroleum for seventeen percent of its energy consumption, its heavy reliance on coal is what makes it a greenhouse giant. By 1990, when China emitted 580 million tons of carbon dioxide a year, it had already surpassed the former Soviet Union as the world's second largest producer of greenhouse gases and trailed only the United States. With its immense coal reserves (over 165 billion tons), huge population, and booming economic modernization program, China will at least double and perhaps triple its greenhouse emissions by 2025, overtaking the United States and profoundly affecting the global struggle against climate change. According to the scientists of the IPCC, greenhouse gas emissions must decline by fifty to seventy percent to stabilize current concentrations of greenhouse gases. If China's carbon dioxide emissions grow as projected between 1990 and 2025, that growth alone would actually increase global emissions by seventeen percent.

Without a radical shift in policies elsewhere in the world, such an increase would doom efforts to achieve the IPCC's target. It would accelerate the global warming already under way and plunge the world into potentially catastrophic territory. And China itself would by no means be immune. A study done by China's National Environmental Protection Agency (NFPA), the World Bank, and the UN Development Program concluded that a doubling of global carbon dioxide concentrations would have the following impacts on China: storms and typhoons would become more extreme and frequent; much of China's coastline, including the economic powerhouses of Shanghai and Guangdong province, would face severe flooding (an area the size of Portugal would be inundated and an estimated sixty-seven million people displaced); agricultural production would be affected most strongly of all, with increased drought and soil erosion lowering average yields of wheat, rice, and cotton; livestock and fish production would also decline. Again, for a country already straining to feed itself, such setbacks could be disastrous.

[. . .]

Coal is the inescapable fact at the center of the Chinese environmental crisis, and there is no easy (or even not so easy) remedy. Although China has taken many steps to mitigate the damages caused by burning coal, its options are limited by a scarcity of everything from water to alternative fuels to investment capital. For example, moving large factories out of city centers, as has been done in Beijing and Chongqing, does not reduce overall air pollution, but it does make the air breathed by city residents less debilitating. More effective, however, would be to "wash" the coal that factories and power plants use before it is burned. Washing, a chemical process routine in the United States and Europe, is especially important for the high-sulfur coal common in China. Nevertheless, only twenty percent of China's coal gets washed, partly because China is desperately short of water; its per capita supply is less than one-third of the world average. Lack of capital is equally problematic, according to Yu Yuefeng of the National People's Congress. "We know the coal should be washed first, but the investment needed to establish coal washing facilities is really enormous," said Yu. "We would need dozens of billions of yuan to achieve this. The Coal Ministry cannot afford it, and since most coal mines are also in poor economic shape, neither can they. So it will be a while before we can increase this twenty percent very much."

The same economic calculus undermines the prevention of acid rain. "Only a very small number of Chinese power plants have sulfur dioxide scrubbers installed, which is a really serious problem," said Wang Wenxing of the Chinese Research Academy of Environmental Sciences. "But China lacks capital. When setting up a power plant, fifteen to thirty percent of the investment funds should be spent on environmental technologies like scrubbers. So China has a choice between building four plants that emit sulfur dioxide and other pollutants or three plants that do not."

Not all environmental technologies are so costly. Perhaps the most important are electrostatic precipitators, which remove particulates from coal smoke yet add only one percent to a power plant's capital cost. Such precipitators have been required on all large Chinese power plants since 1985; they reduce emissions of the largest particulates by ninety percent. Unfortunately, the most recent research indicates that precipitators are useless against the smallest, most harmful types of TSP. "The most dangerous particulates are those less than 2.5 micrometers in diameter," said Wei Fusheng, deputy director of China's National Environmental Monitoring Center, "and neither we nor anyone else in the world has the technology to eliminate them." Wei was among the scientists who served on a long-term research project sponsored by the U.S. Environmental Protection Agency that studied the health effects of air pollution in four Chinese cities: Guangdong, Wuhan, Chongqing, and Lanzhou. He explained that smaller particulates were the most dangerous for two reasons: "First, they get attached to heavy metals like oxidized mercury, lead, and other organic compounds that remain in the atmosphere for a long time. Second, because the particulates are so small, they are not blocked by our bodies' natural defense systems. So they can enter directly into the lungs and even the bloodstream."

Yet China's reliance on coal and its yearning for economic growth are so entrenched that even a health expert like Wei did not oppose doubling the nation's production of coal. Instead, he argued, "we must reduce the health effects of coal burning while we increase the production of coal." In addition to advocating the washing of coal, Wei suggested locating power plants at the mouth of coal mines. "In Beijing, the environmental carrying capacity of the city is reaching its maximum, so no more pollution should be allowed there," he said. "But we can set up a power plant [hundreds of miles away] in Inner Mongolia, next to the coal mine, and transmit the power from there to Beijing." This technical fix would create other problems, however. Where would the vast quantities of water required for the power plants' cooling towers be found in such an arid region? And not even a remote geographic location could neutralize the acid rain and greenhouse gases produced by coal-fired power plants.

So what is to be done? From an environmental and public health standpoint, the best idea is simply to use less coal in the first place. Many Chinese might regard this as a recipe for economic stagnation, but in fact the nation's recent history suggests otherwise. During the 1980s, as part of the economic reform program championed by Deng Xiaoping, China sharply reduced its rate of coal consumption by increasing its energy efficiency. Because the energy intensity of China's economy (i.e., the amount of energy used per unit of GNP produced) fell by thirty percent during the 1980s, the economy was able to grow an average of nine percent a year even as energy consumption grew by only five percent a year.

This remarkable accomplishment was the result of deliberate government policy, implemented in response to the global oil crisis of 1979. Energy planners realized that, in an era of rising oil prices and uncertain supply, China could not hope to produce or buy enough energy to achieve the quadrupling of the Chinese economy by 2000 that Deng Xiaoping wanted; the only option was to use energy more efficiently. Toward that end, economic restructuring shifted the emphasis of China's economy from heavy to light industry. Subsidies and price controls that encouraged waste by making energy appear cheaper than it was were reduced. Introduction of more advanced boilers, fans, and pumps in such core industries as electricity, steel, chemicals, concrete, and fertilizer further increased energy efficiency. More visible perhaps to the average Chinese was the phaseout of honeycomb home stoves—the least efficient, most polluting means of burning coal. This phaseout was pursued in concert with the demolition of vast tracts of *hutongs*, the traditional, one-story courtyard dwellings of urban China, where residents invariably used home stoves for cooking and heating. The *hutongs* were replaced by high-rise apartment buildings that relied on central heating from industrial-sized boilers.

As China prepares for the twenty-first century, efficiency remains the key to its energy future, the one bright spot in an otherwise gloomy picture. Because China began from such a low level of efficiency, there is still a long way to go—which, ironically, is good news for the environment. If the most efficient equipment and processes currently available on the world market were installed throughout China's energy system, the country's energy consumption could be cut by forty to fifty percent, according to studies conducted by Zhou Dadi and his colleagues at the Beijing Energy Efficiency Center. For example, by introducing the most advanced technology for steelmaking and iron smelting at a single large factory in Baoshan, the government reduced the energy intensity of the entire steel industry by ten percent. Meanwhile, experts at the Lawrence Berkeley Laboratories in California have been working with Chinese counterparts to develop refrigerators that would consume half as much energy as today's models. "The Chinese now manufacture more refrigerators than anybody in the world, ten million a year," said Mark

Levine of Berkeley, adding that transforming the Chinese refrigerator industry "would make a major difference" to China and global greenhouse gas emissions. To encourage such reforms, the World Bank has been helping to create energy management corporations that would spread the word about efficiency opportunities within China. The corporations will approach factory managers with a simple message, said the bank's Robert Taylor: "Managing your factory in an energy efficient way will increase your bottom line. And if you don't believe it, here are some case studies that prove it."

Nevertheless, despite the enormous potential of efficiency improvements, the thrust of Chinese energy policy remains the expansion of supply. Although the government directs six to ten percent of its energy budget to efficiency—much more than most other countries, especially poor ones—it invests ten times that much in expanding supply. The emerging private sector is similarly inclined.

Smitten with a capitalism they do not fully understand, many in government think that efficiency no longer needs government support now that prices better reflect economic realities. "Some people in China assume the market will take care of everything," said one top Western consultant. "They are holier than the pope!" In truth, however, since markets rarely internalize the true environmental costs of a given action, their price signals often discourage responsible behavior. For example, a provincial government in China could choose between investing U.S. $1,000 per kilowatt for the standard Chinese power plant, which operates at thirty to forty-five percent efficiency, or U.S. $1,400 for an American-made plant that delivers both forty-five percent efficiency and a ninety percent reduction in sulfur dioxide and nitrogen oxide emissions. Obviously, the American plant is environmentally superior, and it would probably save money in the long run through lower fuel bills as well. But most investors would choose the cheaper Chinese power plant because pervasive capital shortages force them to focus on the short term.

The shortage of capital is most decisive for the smallest end users, who happen to burn more than fifty percent of the coal consumed in China. Particularly in need of overhaul are the nation's estimated 430,000 industrial boilers, many of which were built in the 1950s on the basis of designs from the 1920s. However, "lack of capital means that up-front costs are far more important to the managers who make these investment decisions than lifetime costs are," said Jonathan Sinton, an analyst at the Lawrence Berkeley Laboratories who has interviewed many of these managers. "So instead of replacing what they have, they use the scarce capital to get a new piece of equipment, and they probably don't spend the extra money needed to get environmental features for that equipment."

What about supply-side alternatives to coal? Solar, wind, and nuclear energy each supply less than one percent of total electricity demand. Some planners envision a bigger future role for nuclear, but its high costs and safety problems argue against this. Chinese officials did announce a $2 billion reactor purchase from Russia in 1997, a tempting of fate if ever there was one. Not only was China buying the most accident-prone reactor technology in the world, the reactor would be sited on China's southern coast, amid perhaps the densest concentration of humans on earth. So, what about other options? Hydropower delivers twenty-four percent of China's electricity, and total installed capacity is expected to quadruple by 2020. However, because electricity makes up only fifteen percent of China's total energy use, even hydropower is destined to remain a small share of the overall energy mix. To lower coal consumption an appreciable amount, the heating and industrial sectors have to be tackled. One option under discussion is to expand exploration for natural gas in western China and perhaps even import gas from Russia and Turkmenistan. The most optimistic forecasts hold that natural gas could cover ten percent of China's total energy consumption by 2005—an important step in the right direction, but a relatively small one.

Coal therefore seems destined to remain a large and growing source of energy for China well into the twenty-first century, despite the ominous implications this carries for pollution in China and climate change for the planet as a whole. Bill Clinton has said that he told Chinese president Jiang Zemin in their first meeting, in November 1995, that the biggest security threat China posed to the United States was related not to nuclear weapons or trade agreements but the environment. Specifically, Clinton feared that China would copy America's bad example while developing its economy and end up causing terrible air pollution and global warming. Clinton proudly claimed that he could tell Jiang "had never thought about it just like that." No doubt. Jiang was probably wondering whether Clinton could possibly be serious. If ever there was a nonissue for China's leaders, global warming is it.

"Global warming is not on our agenda," a senior official of the Chongqing Environmental Protection Bureau said with a dismissive wave of his hand when I asked about his agency's strategies to reduce carbon dioxide emissions. As if to underscore his contempt for the issue, the official added an assertion he had to know was false—"All the pollution produced in Chongqing is landing here in Chongqing, so it's not a global problem"— before he declared, "We can't start worrying about carbon dioxide until we solve the sulfur dioxide problem." The official and his counterparts throughout China consider sulfur dioxide more urgent because acid rain is landing on them and causing tangible damage today, while carbon dioxide emissions threaten merely potential, far-off, worldwide damage.

Short-sighted? Yes, but understandable. I arrived in China eager to investigate the climate change issue, but I almost forgot to raise the point during some interviews. When one is inhaling appallingly polluted air for weeks on end, one tends to focus the questions on that.

China has little patience with Western finger-pointing on the climate change issue, regarding it as a cynical means of constraining China's economic development. That is a paranoid view, but it contains a kernel of truth. For all its nuclear weapons, grand ambitions, and expensively dressed businesspeople wielding mobile phones, China remains a very poor country. On a per capita basis, it consumes barely ten percent as much energy as the United States. It is the rich nations whose earlier industrialization has already condemned the world to climate change, argue the Chinese, so why should China be held back? Should not the right to emit greenhouse gases be shared equitably among the world's peoples? To the Chinese, global warming is a rich man's issue, and if he wants China to do something about it, he will have to pay for it. As one Western consultant with regular access to senior Chinese officials put it, "They know very well they can hold the world for ransom…and whenever they can extract concessions, they will."

"The Americans say China is the straw that breaks the camel's back on greenhouse gas emissions," commented Zhou Dadi, a self-described insider on China's climate change policies. "But we say, 'Why don't you take some of your heavy load off the camel first?' If the camel belongs to America, fine, we'll walk. But the camel does not belong to America…. China will insist on the per capita principle [of distributing emissions rights]. What else are we supposed to do? Go back to no heat in winter? Impossible."

Zhou Dadi, I should make clear, is no party hack. He is fully aware of the prospects for global climate change and is doing what he can to prevent it. The traditional communist approach to energy planning, in both the Soviet Union and China, focused on expanding supply through mega-projects like the Three Gorges dam. By contrast, Zhou is an enthusiastic proponent of increasing energy efficiency, which he regards as the smarter and cheaper alternative. As we talked in the lobby of a Beijing hotel one December afternoon, it became clear that Zhou also had personal reasons to reject the old ways. Like so many people of his generation (Zhou was fifty), he had lost some of the best years of his life to the Cultural Revolution. As a young man, he had studied physics at Tsinghua University, the country's premier science university, but upon graduation, he and his classmates were denied formal degrees, which were seen as signs of bourgeois hierarchy. Instead, Zhou was ordered to spend the next nine years in an electronics factory, repairing machinery. Not until the 1980s could he begin reclaiming his professional life by returning to Tsinghua to complete a master's degree in environmental engineering. Despite these hardships, Zhou's loyalty to his country (if not to the Communist Party) remained strong, and—again like many of his peers—the stolen years seemed only to convince him not to squander any more time now.

"China is not like Africa, you know, some remote place that's never been developed," Zhou told me, explaining China's goal of becoming what he called "a middle-class country" like France and Japan. "We used to be the most developed country in the world. Now, after many decades of turbulence, civil war, revolution, political instability, and other difficulties, we finally have the chance to develop the country again. And we will not lose that chance."

*In the United States at the start of the new millenium, climate skepticism was on the rise, fueled by the ascension of the George W. Bush administration, which had deep roots in the oil industry. Bush repudiated any effort at carbon control soon after taking office, and underlings in his administration routinely pressured federal scientists to alter their reports or stop talking to reporters. On Capitol Hill, no one captured the spirit of the moment more fully than Oklahoma Senator James Inhofe, who famously called global warming a "hoax" and used his position as the chair of the relevant Senate committee to block any legislation. The recipient of massive amounts of campaign funding from the oil and gas industries that dominated the Oklahoma economy, Inhofe used his position to put forward an ongoing critique of mainstream science. The reader should know that virtually all the scientific attacks the Senator makes have been answered over and over at least as far back as Gelbspan's book (see page 105); still, it's important to record the lines of attack climate deniers have used over and over.*

# The Science of Climate Change: Senate Floor Statement

James M. Inhofe
*2003*

As chairman of the Committee on Environment and Public Works, I have a profound responsibility, because the decisions of the committee have wide-reaching impacts, influencing the health and security of every American.

That's why I established three guiding principles for all committee work: it should rely on the most objective science; it should consider costs on businesses and consumers; and the bureaucracy should serve, not rule, the people.

Without these principles, we cannot make effective public policy decisions. They are necessary to both improve the environment and encourage economic growth and prosperity.

One very critical element to our success as policymakers is how we use science. That is especially true for environmental policy, which relies very heavily on science. I have insisted that federal agencies use the best, nonpolitical science to drive decision making. Strangely, I have been harshly criticized for taking this stance. To the environmental extremists, my insistence on sound science is outrageous.

For them, a "pro-environment" philosophy can only mean top-down, command-and-control rules dictated by bureaucrats. Science is irrelevant—instead, for extremists, politics and power are the motivating forces for making public policy.

But if the relationship between public policy and science is distorted for political ends, the result is flawed policy that hurts the environment, the economy, and the people we serve.

Sadly, that's true of the current debate over many environmental issues. Too often emotion, stoked by irresponsible rhetoric, rather than facts based on objective science, shapes the contours of environmental policy.

A rather telling example of this arose during President Bush's first days in office, when emotionalism overwhelmed science in the debate over arsenic standards in drinking water. Environmental groups, including the Sierra Club and the Natural Resources Defense Council, vilified President Bush for "poisoning" children because he questioned the scientific basis of a regulation implemented in the final days of the Clinton Administration.

The debate featured television ads, financed by environmental groups, of children asking for another glass of arsenic-laden water. The science underlying the standard, which was flimsy at best, was hardly mentioned or held up to any scrutiny.

The Senate went through a similar scare back in 1992. That year some members seized on data from NASA suggesting that an ozone hole was developing in the Northern Hemisphere. The Senate then rushed into panic, ramming through, by a ninety-six-to-zero vote, an accelerated ban on certain chlorofluorocarbon refrigerants. Only two weeks later NASA produced new data showing that their initial finding was a gross exaggeration, and the ozone hole never appeared.

The issue of catastrophic global warming, which I would like to speak about today, fits perfectly into this mold. Much of the debate over global warming is predicated on fear, rather than science. Global warming alarmists see a future plagued by catastrophic flooding, war, terrorism, economic dislocations, droughts, crop failures, mosquito-borne diseases, and harsh weather—all caused by man-made greenhouse gas emissions.

Hans Blix, chief U.N. weapons inspector, sounded both ridiculous and alarmist when he said in March, "I'm more worried about global warming than I am of any major military conflict."

Science writer David Appell, who has written for such publications as the *New Scientist* and *Scientific American*, parroted Blix when he said global warming would "threaten fundamental food and water sources. It would lead to displacement of billions of people and huge waves of refugees, spawn terrorism and topple governments, spread disease across the globe."

Appell's next point deserves special emphasis, because it demonstrates the sheer lunacy of environmental extremists: "[Global warming] would be chaos by any measure, far greater even than the sum total of chaos of the global wars of the twentieth century, and so in this sense Blix is right to be concerned. Sounds like a weapon of mass destruction to me."

No wonder the late political scientist Aaron Wildavsky called global warming alarmism the "mother of all environmental scares."

Appell and Blix sound very much like those who warned us in the 1970s that the planet was headed for a catastrophic global cooling. On April 28, 1975, *Newsweek* printed an article titled "The Cooling World," in which the magazine warned: "There are ominous signs that the earth's weather patterns have begun to change dramatically and that these changes may portend a drastic decline in food production—with serious political implications for just about every nation on earth."

In a similar refrain, *Time* magazine for June 24, 1974, declared: "However widely the weather varies from place to place and time to time, when meteorologists take an average of temperatures around the globe they find that the atmosphere has been growing gradually cooler for the past three decades."

In 1974 the National Science Board, the governing body of the National Science Foundation, stated: "During the last twenty to thirty years, world temperature has fallen, irregularly at first but more sharply over the last decade." Two years earlier, the board had observed: "Judging from the record of the past inter-glacial ages, the present time of high temperatures should be drawing to an end...leading into the next glacial age."

How quickly things change. Fear of the coming ice age is old hat, but fear that man-made greenhouse gases are causing temperatures to rise to harmful levels is in vogue. Alarmists brazenly assert that this phenomenon is fact, and that the science of climate change is "settled."

To cite just one example, Ian Bowles, former senior science director on environmental issues for the Clinton National Security Council, said in the April 22, 2001, edition of the *Boston Globe*: "The basic link between carbon emissions, accumulation of greenhouse gases in the atmosphere, and the phenomenon of climate change is not seriously disputed in the scientific community."

But in fact the issue is far from settled, and indeed is seriously disputed. I would like to submit at the end of my remarks a July 8 editorial by former Carter administration Energy Secretary James Schlesinger on the science of climate change. In that editorial, Dr. Schlesinger takes issue with alarmists who assert there is a scientific consensus supporting their views.

"There is an idea among the public that the science is settled," Dr. Schlesinger wrote. "[T]hat remains far from the truth."

Today, even saying there is scientific disagreement over global warming is itself controversial. But anyone who pays even cursory attention to the issue understands that scientists vigorously disagree over whether human activities are responsible for global warming, or whether those activities will precipitate natural disasters.

I would submit, furthermore, that not only is there a debate, but the debate is shifting away from those who subscribe to global warming alarmism. After studying the issue over the last several years, I believe that the balance of the evidence offers strong proof that natural variability is the overwhelming factor influencing climate.

It's also important to question whether global warming is even a problem for human existence. Thus far no one has seriously demonstrated any scientific proof that increased global temperatures would lead to the catastrophes predicted by alarmists. In fact, it appears that just the opposite is true: that increases in global temperatures may have a beneficial effect on how we live our lives.

For these reasons I would like to discuss an important body of scientific research that refutes the anthropogenic theory of catastrophic global warming. I believe this research offers compelling proof that human activities have little impact on climate.

This research, well documented in the scientific literature, directly challenges the environmental worldview of the media, so it typically doesn't receive proper attention and discussion. Certain members of the media would rather level personal attacks on scientists who question "accepted" global warming theories than engage on the science.

This is an unfortunate artifact of the debate—the relentless increase in personal attacks on certain members of the scientific community who question so-called conventional wisdom.

I believe it is extremely important for the future of this country that the facts and the science get a fair hearing. Without proper knowledge and understanding, alarmists will scare the country into enacting its ultimate goal: making energy suppression, in the form of harmful mandatory restrictions on carbon dioxide and other greenhouse emissions, the official policy of the United States.

Such a policy would induce serious economic harm, especially for low-income and minority populations. Energy suppression, as official government and nonpartisan private analyses have amply confirmed, means higher prices for food, medical care, and electricity, as well as massive job losses and drastic reductions in gross domestic product, all the while providing virtually no environmental benefit. In other words: a raw deal for the American people and a crisis for the poor.

[. . .]

*IPCC Assessment Reports*

In 1992, several nations from around the globe gathered in Rio de Janeiro for the United Nations Framework Convention on Climate Change. The meeting was premised on the concern that global warming was becoming a problem. The U.S., along with many others, signed the Framework Convention, committing them to making voluntary reductions in greenhouse gases.

Over time, it became clear that signatories were not achieving their reduction targets as stipulated under Rio. This realization led to the Kyoto Protocol in 1997, which was an amendment to the Framework Convention, and which prescribed mandatory reductions only for developed nations. (By the way, leaving out developing nations was an explicit violation of Byrd-Hagel.)

The science of Kyoto is based on the "Assessment Reports" conducted by the Intergovernmental Panel on Climate Change, or IPCC. Over the last thirteen years, the IPCC has published three assessments, with each one over time growing more and more alarmist.

The first IPCC Assessment Report in 1990 found that the climate record of the past century was "broadly consistent" with the changes in Earth's surface temperature, as calculated by climate models that incorporated the observed increase in greenhouse gases.

This conclusion, however, appears suspect considering the climate cooled between 1940 and 1975, just as industrial activity grew rapidly after World War II. It has been difficult to reconcile this cooling with the observed increase in greenhouse gases.

After its initial publication, the IPCC's second Assessment Report in 1995 attracted widespread international attention, particularly among scientists who believed that human activities were causing global warming. In their view, the report provided the proverbial smoking gun.

The most widely cited phrase from the report—actually, it came from the report summary, as few in the media actually read the entire report—was that "the balance of the evidence suggests a discernible human influence on global climate." This of course is so vague that it's essentially meaningless.

What do they mean by "suggests"? And, for that matter, what, in this particular context, does "discernible" mean? How much human influence is discernible? Is it a positive or negative influence? Where is the precise scientific quantification?

Unfortunately the media created the impression that man-induced global warming was fact. On August 10, 1995, the *New York Times* published an article titled "Experts Confirm Human Role in Global Warming." According to the *Times* account, the IPCC showed that global warming "is unlikely to be entirely due to natural causes."

Of course, when parsed, this account means fairly little. Not entirely due to natural causes? Well, how much, then? One percent? Twenty percent? Eighty-five percent?

The IPCC report was replete with caveats and qualifications, providing little evidence to support anthropogenic theories of global warming. The preceding paragraph in which the "balance of evidence" quote appears makes exactly that point.

It reads: "Our ability to quantify the human influence on global climate is currently limited because the expected signal is still emerging from the noise of natural variability, and because there are uncertainties in key factors. These include the magnitude and patterns of long-term variability and the time evolving pattern of forcing by, and response to, changes in concentrations of greenhouse gases and aerosols, and land surface changes."

Moreover, the IPCC report was quite explicit about the uncertainties surrounding a link between human actions and global warming. "Although these global mean results suggest that there is some anthropogenic component in the observed temperature record, they cannot be considered compelling evidence of a clear cause-and-effect link between anthropogenic forcing and changes in the Earth's surface temperature."

Remember, the IPCC provides the scientific basis for the alarmists' conclusions about global warming. But even the IPCC is saying that their own science cannot be considered compelling evidence.

Dr. John Christy, professor of Atmospheric Science and Director of the Earth System Science Center at the University of Alabama in Huntsville, and a key contributor to the 1995 IPCC report, participated with the lead authors in the drafting sessions, and in the detailed review of the scientific text. He wrote in the Montgomery *Advertiser* on February 22, 1998, that much of what passes for common knowledge in the press regarding climate change is "inaccurate, incomplete or viewed out of context."

Many of the misconceptions about climate change, Christy contends, originated from the IPCC's six-page executive summary. It was the most widely read and quoted of the three documents published by the IPCC's Working Group, but, Christy said—and this point is crucial—it had the "least input from scientists and the greatest input from non-scientists."

## IPCC Releases Third Assessment on Climate Change

Five years later, the IPCC was back again, this time with the Third Assessment Report on Climate Change. In October of 2000, the IPCC "Summary for Policymakers" was leaked to the media, which once again accepted the IPCC's conclusions as fact.

Based on the summary, the *Washington Post* wrote on October 30, "The consensus on global warming keeps strengthening." In a similar vein, the *New York Times* confidently declared on October 28, "The international panel of climate scientists considered the most authoritative voice on global warming has now concluded that mankind's contribution to the problem is greater than originally believed."

Note again, look at how these accounts are couched: they are worded to maximize the fear factor. But upon closer inspection, it's clear that such statements have no compelling intellectual content. "Greater than originally believed"? What is the baseline from which the *Times* makes such a judgment? Is it 0.01 percent, or twenty-five percent? And how much is greater? Double? Triple? An order of magnitude greater?

Such reporting prompted testimony by Dr. Richard Lindzen before the Committee on Environment and Public Works, the committee I now chair, in May of 2001. Lindzen said, "Nearly all reading and coverage of the IPCC is restricted to the highly publicized Summaries for Policymakers, which are written by representatives from governments, NGOs, and business; the full reports, written by participating scientists, are largely ignored."

As it turned out, the Policymaker's Summary was politicized and radically differed from an earlier draft. For example, the draft concluded the following concerning the driving causes of climate change:

> From the body of evidence since IPCC (1996), we conclude that there has been a discernible human influence on global climate. Studies are beginning to separate the contributions to observed climate change attributable to individual external influences, both anthropogenic and natural. This work suggests that anthropogenic greenhouse gases are a substantial contributor to the observed warming, especially over the past thirty years. However, the accuracy of these estimates continues to be limited by uncertainties in estimates of internal variability, natural and anthropogenic forcing, and the climate response to external forcing.

The final version looks quite different, and concluded instead: "In the light of new evidence and taking into account the remaining uncertainties, most of the observed warming over the last fifty years is likely to have been due to the increase in greenhouse gas concentrations."

This kind of distortion was not unintentional, as Dr. Lindzen explained before the EPW Committee. He said, "I personally witnessed coauthors forced to assert their 'green' credentials in defense of their statements."

In short, some parts of the IPCC process resembled a Soviet-style trial, in which the facts are predetermined, and ideological purity trumps technical and scientific rigor.

The predictions in the summary went far beyond those in the IPCC's 1995 report. In the Second Assessment, the IPCC predicted that the earth could warm by 1 to 3.5 degrees Celsius by the year 2100. The "best estimate" was a two-degree-Celsius warming by 2100. Both are highly questionable at best.

In the Third Assessment, the IPCC dramatically increased that estimate to a range of 1.4 to 5.8 degrees Celsius, even though no new evidence had come to light to justify such a dramatic change.

In fact, the IPCC's median projected warming actually declined from 1990 to 1995. The IPCC 1990 initial estimate was 3.2 degrees Celsius, then the IPCC revised 1992 estimate was 2.6 degrees Celsius, followed by the IPCC revised 1995 estimate of 2.0 degrees Celsius.

What changed? As it turned out, the new prediction was based on faulty, politically charged assumptions about trends in population growth, economic growth, and fossil fuel use.

The extreme-case scenario of a 5.8-degree warming, for instance, rests on an assumption that the whole world will raise its level of economic activity and per capita energy use to that of the United States, and that energy use will be carbon-intensive. This scenario is simply ludicrous. This essentially contradicts the experience of the industrialized world over the last thirty years. Yet the 5.8-degree figure featured prominently in news stories because it produced the biggest fear effect.

Moreover, when regional climate models, of the kind relied upon by the IPCC, attempt to incorporate such factors as population growth, "the details of future climate recede toward unintelligibility," according to Jerry Mahlman, Director of NOAA's Geophysical Fluid Dynamics Laboratory.

Even Dr. Stephen Schneider, an outspoken believer in catastrophic global warming, criticized the IPCC's assumptions in the journal *Nature* on May 3, 2001. In his article, Schneider asks, "How likely is it that the world will get 6 degrees C hotter by 2100?" That, he said, "depends on the likelihood of the assumptions underlying the projections."

The assumptions, he wrote, are " 'storylines' about future worlds from which population, affluence and technology drivers could be inferred." These storylines, he wrote, "gave rise to radically different families of emission profiles up to 2100—from below current $CO_2$ emissions to five times current emissions."

Schneider says that he "strongly argued at the time that policy analysts needed probability estimates to assess the seriousness of the implied impacts." In other words, how likely is it that temperatures would go up by 5.8 degrees Celsius, or 1.4 degrees Celsius, which represent the IPCC's respective upper and lower bounds?

But as Schneider wrote, the group drafting the IPCC report decided to express "no preference" for each temperature scenario.

In effect, this created the assumption that the higher bound of 5.8 degrees Celsius appeared to be just as likely as the lower of 1.4 degrees Celsius. "But this inference would be incorrect," said

Schneider, "because uncertainties compound through a series of modeling steps."

Keep in mind here that Schneider is on the side of the alarmists. Schneider's own calculations, which cast serious doubt on the IPCC's extreme prediction, broadly agree with an MIT study published in April of 2001. It found that there is a "far less" than one percent chance that temperatures would rise to 5.8 degrees Celsius or higher, while there is a seventeen percent chance the temperature rise would be lower than 1.4 degrees.

That point bears repeating: even true believers think the lower number is seventeen times more likely to be right than the higher number. Moreover, even if the earth's temperature increases by 1.4 degrees Celsius, does it really matter? The IPCC doesn't offer any credible science to explain what would happen.

Gerald North of Texas A&M University in College Station agrees that the IPCC's predictions are baseless, in part because climate models are highly imperfect instruments. As he said after the IPCC report came out: "It's extremely hard to tell whether the models have improved" since the last IPCC report. "The uncertainties are large." Similarly, Peter Stone, an MIT climate modeler, said in reference to the IPCC, "The major [climate prediction] uncertainties have not been reduced at all."

Dr. David Wojick, an expert in climate science, recently wrote in Canada's *National Post*, "The computer models cannot...decide among the variable drivers, like solar versus lunar change, or chaos versus ocean circulation versus greenhouse gas increases. Unless and until they can explain these things, the models cannot be taken seriously as a basis for public policy."

In short, these general circulation models, or GCMs as they're known, create simulations that must track over 5 million parameters. These simulations require accurate information on two natural greenhouse gas factors—water vapor and clouds—whose effects scientists still do not understand.

Even the IPCC conceded as much: "The single largest uncertainty in determining the climate sensitivity to either natural or anthropogenic changes are clouds and their effects on radiation and their role in the hydrological cycle...at the present time, weaknesses in the parameterization of cloud formation and dissipation are probably the main impediment to improvements in the simulation of cloud effects on climate."

Because of these and other uncertainties, climate modelers from four separate climate modeling centers wrote in the October 2000 edition of *Nature* that, "Forecasts of climate change are inevitably uncertain." They go on to explain that, "A basic problem with all such predictions to date has been the difficulty of providing any systematic estimate of uncertainty," a problem that stems from the fact that "These [climate] models do not necessarily span the full range of known climate system behavior."

Again, to reiterate in plain English, this means the models do not account for key variables that influence the climate system.

Despite this, the alarmists continue to use these models and all the other flimsy evidence I've cited to support their theories of man-made global warming.

### The Twentieth Century: Satellite Data, Weather Balloons, $CO_2$, and Glaciers

Now I want to turn to temperature trends in the twentieth century. GCMs predict that rising atmospheric $CO_2$ concentrations will cause temperatures in the troposphere, the layer from 5,000 to

30,000 feet, to rise faster than surface temperatures—a critical fact supporting the alarmist hypothesis.

But in fact, there is no meaningful warming trend in the troposphere, and weather satellites, widely considered the most accurate measure of global temperatures, have confirmed this.

To illustrate this point, just think about a greenhouse. The glass panes let sunlight in but prevent it from escaping. The greenhouse then warms from the top down. As is clear from the science, this simply is not happening in the atmosphere.

Satellite measurements are validated independently by measurements from NOAA balloon radiosonde instruments, whose records extend back over forty years.

If you look at this chart of balloon data extremists will tell you that warming is occurring, but if you look more closely you see that temperature in 1955 was higher than temperature in 2000.

A recent detailed comparison of atmospheric temperature data gathered by satellites with widely used data gathered by weather balloons corroborates both the accuracy of the satellite data and the rate of global warming seen in that data.

Using NOAA satellite readings of temperatures in the lower atmosphere, scientists at the University of Alabama in Huntsville produced a dataset that shows global atmospheric warming at the rate of about 0.07 degrees Celsius (about 0.13 degrees Fahrenheit) per decade since November 1978.

"That works out to a global warming trend of about one and a quarter degrees Fahrenheit over one hundred years," said Dr. John Christy, who compiled the comparison data. Christy concedes that such a trend "is probably due in part to human influences," but adds that "it's substantially less than the warming forecast by most climate models, and"—here is the key point—"it isn't entirely out of the range of climate change we might expect from natural causes."

To reiterate: the best data collected from satellites validated by balloons to test the hypothesis of a human-induced global warming from the release of $CO_2$ into the atmosphere shows no meaningful trend of increasing temperatures, even as the climate models exaggerated the warmth that ought to have occurred from a build-up in $CO_2$.

Some critics of satellite measurements contend that they don't square with the ground-based temperature record. But some of this difference is due to the so-called "urban heat island effect." This occurs when concrete and asphalt in cities absorb—rather than reflect—the sun's heat, causing surface temperatures and overall ambient temperatures to rise. Scientists have shown that this strongly influences the surface-based temperature record.

In a paper published in the *Bulletin of the American Meteorological Society* in 1989, Dr. Thomas R. Karl, senior scientist at the National Climate Data Center, corrected the U.S. surface temperatures for the urban heat-island effect and found that there has been a downward temperature trend since 1940. This suggests a strong warming bias in the surface-based temperature record.

Even the IPCC finds that the urban heat island effect is significant. According to the IPCC's calculations, the effect could account for up to 0.12 degrees Celsius of the twentieth-century temperature rise, one-fifth of the total observed.

When we look at the twentieth century as a whole, we see some distinct phases that question anthropogenic theories of global warming. First, a strong warming trend of about 0.5 degrees Celsius began in the late nineteenth century and peaked around 1940. Next, the temperature decreased from 1940 until the late 1970s.

Why is that decrease significant? Because about eighty percent of the carbon dioxide from human activities was added to the air after 1940, meaning the early twentieth century warming trend had to be largely natural.

Scientists from the Scripps Institution for Oceanography confirmed this phenomenon in the March 12, 1999, issue of the journal *Science*. They addressed the proverbial "chicken-and-egg" question of climate science, namely, when the earth shifts from glacial to warm periods, which comes first: an increase in atmospheric carbon dioxide levels, or an increase in global temperature?

The team concluded that the temperature rise comes first, followed by a carbon dioxide boost 400 to 1,000 years later. This contradicts everything alarmists have been saying about man-made global warming in the twentieth century.

Now we can even go back 400,000 years and see this phenomenon occurring, as this chart clearly shows.

Yet the doomsayers, undeterred by these facts, just won't quit. In February and March of 2002, the *New York Times* and the *Washington Post*, among others, reported on the collapse of the Larsen B ice shelf in the Antarctic Peninsula, causing quite a stir in the media, and providing alarmists with more propaganda to scare the public.

Although there was no link to global warming, the *Times* couldn't help but make that suggestion in its March 20 edition. "While it is too soon to say whether the changes there are related to a buildup of the 'greenhouse' gas emissions that scientists believe are warming the planet, many experts said it was getting harder to find any other explanation."

The *Times*, however, simply ignored a recent study in the journal *Nature*, which found the Antarctic has been cooling since 1966. And another study in *Science* recently found the West Antarctic Ice Sheet has been thickening rather than thinning.

University of Illinois researchers also reported "a net cooling on the Antarctic continent between 1966 and 2000." In some regions, like the McMurdo Dry Valleys, temperatures cooled between 1986 and 1999 by as much as two degrees centigrade per decade.

In perhaps the most devastating critique of glacier alarmism, the American Geophysical Union found that the Arctic was warmer in 1935 than it is now. "Two distinct warming periods from 1920 to 1945, and from 1975 to the present, are clearly evident...compared with the global and hemispheric temperature rise, the high-latitude temperature increase was stronger in the late 1930s to early 1940s than in recent decades."

Again, that bears repeating: eighty percent of the carbon dioxide from human activities was added to the air after 1940—yet the Arctic was warmer in 1935 than it is today.

So, not only is glacier alarmism flawed, but there is no evidence, as shown by measurements from satellites and weather balloons, of any meaningful warming trends in the twentieth century.

## Global Warming Health Risks/Benefits

Even as we discuss whether temperatures will go up or down, we should ask whether global warming would actually produce the catastrophic effects its adherents so confidently predict.

What gets obscured in the global warming debate is the fact that carbon dioxide is not a pollutant. It is necessary for life.

Numerous studies have shown that global warming can actually be beneficial to mankind.

Most plants, especially wheat and rice, grow considerably better when there is more $CO_2$ in the atmosphere. $CO_2$ works like a fertilizer and higher temperatures usually further enhance the $CO_2$ fertilizer effect.

In fact the average crop, according to Dr. John Reilly, of the MIT Joint Program on the Science and Policy of Global Change, is thirty percent higher in a $CO_2$-enhanced world. I want to repeat that: PRODUCTIVITY IS THIRTY PERCENT HIGHER IN A $CO_2$-ENHANCED WORLD. This is not just a matter of opinion, but a well-established phenomenon.

With regard to the impact of global warming on human health, it is assumed that higher temperatures will induce more deaths and massive outbreaks of deadly diseases. In particular, a frequent scare tactic by alarmists is that warmer temperatures will spark malaria outbreaks. Dr. Paul Reiter convincingly debunks this claim in a 2000 study for the Center for Disease Control. As Reiter found, "Until the second half of the twentieth century, malaria was endemic and widespread in many temperate regions"—this next point is critical—"with major epidemics as far north as the Arctic Circle."

Reiter also published a second study in the March 2001 issue of *Environmental Health Perspectives* showing that "despite spectacular cooling [of the Little Ice Age], malaria persisted throughout Europe."

Another myth is that warming increases morbidity rates. This isn't the case, according to Dr. Robert Mendelsohn, an environmental economist from Yale University. Mendelsohn argues that heat-stress deaths are caused by temperature variability and not warming. Those deaths grow in number not as climates warm but as the variability in climate increases.

[. . .]

*The Next Steps*

I am mystified that some in this body, and in the media, blithely assert that the science of global warming is settled—that is, fossil fuel emissions are the principal driving cause of global warming. In a recent letter to me concerning the next EPA administrator, two senators wrote that "the pressing problem of global warming" is now an "established scientific fact," and demanded that the new administrator commit to addressing it.

With all due respect, this statement is baseless, for several reasons. As I outlined in detail above, the evidence is overwhelmingly in favor of those who don't see global warming posing grave harm to the planet and who don't think human beings have significant influence on the climate system.

This leads to another question: why would this body subject the United States to Kyoto-like measures that have no environmental benefits and cause serious harm to the economy? There are several pieces of legislation, including several that have been referred to my committee, that effectively implement Kyoto. From a cursory read of Senate politics, it is my understanding that some of these bills enjoy more than a modicum of support.

I urge my colleagues to reject them, and follow the science to the facts. Reject approaches designed not to solve an environmental problem, but to satisfy the ever-growing demand of environmental groups for money and power and other extremists who simply don't like capitalism, free markets, and freedom.

Climate alarmists see an opportunity here to tax the American people. Consider a July 11 op-ed by J.W. Anderson in the *Washington Post*. In it, Anderson, a former editorial writer for the *Post*, and now a journalist in residence with Resources for the Future, concedes that climate science still confronts uncertainties. But his solution is a fuel tax to prepare for a potentially catastrophic future. Based on the case I've outlined today, such a course of action fits a particular ideological agenda, yet is entirely unwarranted.

It is my fervent hope that Congress will reject prophets of doom who peddle propaganda masquerading as science in the name of saving the planet from catastrophic disaster. I urge my colleagues to put stock in scientists who rely on the best, most objective scientific data and reject fear as a motivating basis for making public policy decisions.

Let me be very clear: alarmists are attempting to enact an agenda of energy suppression that is inconsistent with American values of freedom, prosperity, and environmental progress.

Over the past two hours, I have offered compelling evidence that catastrophic global warming is a hoax. That conclusion is supported by the painstaking work of the nation's top climate scientists.

What have those scientists concluded? The Kyoto Protocol has no environmental benefits; natural variability, not fossil fuel emissions, is the overwhelming factor influencing climate change; satellite data, confirmed by NOAA balloon measurements, confirms that no meaningful warming has occurred over the last century; and climate models predicting dramatic temperature increases over the next one hundred years are flawed and highly imperfect.

*Climate Experts*

I want to recount who these scientists are:

Dr. S. Fred Singer, an atmospheric scientist at the University of Virginia, who served as the first director of the U.S. Weather Satellite Service (which is now in the Department of Commerce) and more recently as a member and vice chairman of the National Advisory Committee on Oceans and Atmosphere (NACOA).

Dr. Tom Wigley, a senior scientist at the National Center for Atmospheric Research, who found that if the Kyoto Protocol were fully implemented by all signatories, it would reduce temperatures by a mere 0.07 degrees Celsius by 2050, and 0.13 degrees Celsius by 2100. What does this mean? Such an amount is so small that ground-based thermometers cannot reliably measure it.

Dr. Richard Lindzen, an MIT scientist and member of the National Academy of Sciences, who has specialized in climate issues for over thirty years.

Jerry Mahlman, director of NOAA's Geophysical Fluid Dynamics Laboratory, who points out that when regional climate models, of the kind relied upon by the IPCC, attempt to incorporate such factors as population growth, "the details of future climate recede toward unintelligibility."

Gerald North of Texas A&M University in College Station, agrees that the IPCC's predictions are baseless, in part because climate models are highly imperfect instruments. As he said after the IPCC report came out: "It's extremely hard to tell whether the models have improved" since the last IPCC report. "The uncertainties are large."

Peter Stone, an MIT climate modeler, said in reference to the IPCC, "The major [climate prediction] uncertainties have not been reduced at all."

Dr. David Wojick, an expert in climate science, who recently wrote in an article in Canada's *National Post*, "The computer models cannot…decide among the variable drivers, like solar versus lunar change, or chaos versus ocean circulation versus greenhouse gas increases. Unless and until they can explain these things, the models cannot be taken seriously as a basis for public policy."

Climate modelers from four separate climate modeling centers who wrote in the October 2000 edition of *Nature* that, "Forecasts of climate change are inevitably uncertain." They go on to explain that, "A basic problem with all such predictions to date has been the difficulty of providing any systematic estimate of uncertainty," a problem that stems from the fact that "These [climate] models do not necessarily span the full range of known climate system behavior."

NASA scientists Roy Spencer and John Christy, whose satellite data, validated independently by measurements from NOAA balloon radiosonde instruments, show that the atmosphere has not warmed as alarmists theorize.

Dr. Thomas R. Karl, senior scientist at the National Climate Data Center, who corrected the U.S. surface temperatures for the urban heat island effect and found that there has been a downward temperature trend since 1940. This suggests a strong warming bias in the surface-based temperature record.

Scientists from the Scripps Institution for Oceanography who concluded that the temperature rise comes first, followed by a carbon dioxide boost 400 to 1,000 years later. This contradicts everything alarmists have been saying about man-made global warming in the twentieth century.

University of Illinois researchers who reported "a net cooling on the Antarctic continent between 1966 and 2000." In some regions, like the McMurdo Dry Valleys, temperatures cooled between 1986 and 1999 by as much as two degrees centigrade per decade.

Dr. Paul Reiter, who convincingly debunks the claim that higher temperatures will induce more deaths and massive outbreaks of deadly diseases in a 2000 study for the Centers for Disease Control.

Dr. David Legates, a renowned professor at the University of Delaware and the world's leading expert in the hydrology of climate.

Over 4,000 scientists, seventy of whom are Nobel Prize winners, who signed the Heidelberg Appeal, which says that no compelling evidence exists to justify controls of anthropogenic greenhouse gas emissions.

I also point to a 1998 recent survey of state climatologists, which reveals that a majority of respondents have serious doubts about whether anthropogenic emissions of greenhouse gases present a serious threat to climate stability.

Drs. Willie Soon and Sallie Baliunas of the Harvard-Smithsonian Center for Astrophysics, who have just completed the most comprehensive review of temperature records ever.

Then there is Dr. Frederick Seitz, a past president of the National Academy of Sciences, and a professor emeritus at Rockefeller University.

Over 17,000 independently verified signers of the Oregon Petition, which reads as follows:

> We urge the United States government to reject the global warming agreement that was written in Kyoto, Japan, in December 1997, and any other similar proposals. The proposed limits on greenhouse gases would harm the

environment, hinder the advance of science and technology, and damage the health and welfare of mankind.

There is no convincing scientific evidence that human release of carbon dioxide, methane, or other greenhouse gases is causing or will, in the foreseeable future, cause catastrophic heating of the earth's atmosphere and disruption of the earth's climate. Moreover, there is substantial scientific evidence that increases in atmospheric carbon dioxide produce many beneficial effects upon the natural plant and animal environments of the earth.

Kenneth Green, D.Env., is chief scientist and director of the Risk and Environment Centre at the Fraser Institute. He most recently wrote *Global Warming: Understanding the Debate*.

George H. Taylor, who is the state climatologist for Oregon, and a faculty member at Oregon State University's College of Oceanic and Atmospheric Sciences, manages the Oregon Climate Service, the state repository of weather and climate information. Mr. Taylor is a member of the American Meteorological Society and is past president of the American Association of State Climatologists.

Pat Michaels is a research professor of environmental sciences at the University of Virginia and visiting scientist with the Marshall Institute in Washington, D.C. He is a past president of the American Association of State Climatologists and was program chair for the Committee on Applied Climatology of the American Meteorological Society. Michaels has authored tests on climate and is a contributing author and reviewer of the United Nations Intergovernmental Panel on Climate Change.

According to *Nature* magazine, Pat Michaels may be the most popular lecturer in the nation on the subject of global warming.

Freeman Dyson, professor of physics at the Institute for Advanced Study, Princeton University, since 1953, is a fellow of the Royal Society, a member of the U.S. National Academy of Science, and has received numerous international awards.

Robert Balling Jr.; professor and director of the Office of Climatology at Arizona State University who received his PhD from the University of Oklahoma, has authored three books on climate.

Professor Chris Essex of the University of Western Ontario and of the Niels Bohr Institute's Orsted Laboratory and the Canadian Climate Center coauthored *Taken by Storm* with Professor Ross McKitrick of the University of Guelph and the Fraser Institute in Vancouver.

Dr. John Reilly, of the MIT Joint Program on the Science and Policy of Global Change, who established the benefits of $CO_2$ on flora.

And many, many others.

Finally, I will return to the words of Dr. Frederick Seitz, a past president of the National Academy of Sciences, and a professor emeritus at Rockefeller University, who compiled the Oregon Petition: "There is no convincing scientific evidence that human release of carbon dioxide, methane, or other greenhouse gasses is causing or will, in the foreseeable future, cause catastrophic heating of the earth's atmosphere and disruption of the earth's climate. Moreover, there is substantial scientific evidence that increases in atmospheric carbon dioxide produce many beneficial effects upon the natural plant and animal environments of the earth."

These are sobering words, which the extremists have chosen to ignore. So what could possibly be the motivation for global warming alarmism? Since I've become chairman of the EPW Committee, it's become pretty clear: fundraising. Environmental extremists rake in millions of dollars, not to solve environmental problems, but to fuel their ever-growing fundraising machines, part of which are financed by federal taxpayers.

So what have we learned from the scientists and economists I've talked about today?

The claim that global warming is caused by man-made emissions is simply untrue and not based on sound science.

$CO_2$ does not cause catastrophic disasters—actually it would be beneficial to our environment and our economy.

Kyoto would impose huge costs on Americans, especially the poor.

The motives for Kyoto are economic, not environmental—that is, proponents favor handicapping the American economy through carbon taxes and more regulations.

With all of the hysteria, all of the fear, all of the phony science, could it be that man-made global warming is the greatest hoax ever perpetrated on the American people? It sure sounds like it.

*Written for popular consumption, Michael Crichton's* State of Fear *probably did more than Inhofe's Senate speeches to swell the ranks of the skeptics. (And indeed Crichton was invited to testify before Congress on several occasions.) The author of a skein of bestsellers, dating back to* The Andromeda Strain, *which made unlikely events into scary yarns, Crichton reversed his technique with this book. It argued (with copious footnotes that interrupted the story) that global warming was a scam perpetrated by environmentalists in order to raise money and gain planetary control.*

# from *State of Fear*

Michael Crichton
2004

STANGFEDLIS
MONDAY, AUGUST 23
3:02 A.M.

Christ, it was cold, George Morton thought, climbing out of the Land Cruiser. The millionaire philanthropist stamped his feet and pulled on gloves, trying to warm himself. It was three o'clock in the morning, and the sky glowed red, with streaks of yellow from the still-visible sun. A bitter wind blew across the Sprengisandur, the rugged, dark plain in the interior of Iceland. Flat gray clouds hung low over the lava that stretched away for miles. The Icelanders loved this place. Morton couldn't see why.

In any case, they had reached their destination: directly ahead lay a huge, crumpled wall of dirt-covered snow and rock, stretching up to the mountains behind. This was Snorrajökul, one tongue of the huge Vatnajökull glacier, the largest ice cap in Europe.

The driver, a graduate student, climbed out and clapped his hands with delight. "Not bad at all! Quite warm! You are lucky, it's a pleasant August night." He was wearing a T-shirt, hiking shorts, and a light vest. Morton was wearing a down vest, a quilted windbreaker, and heavy pants. And he was still cold.

He looked back as the others got out of the backseat. Nicholas Drake, thin and frowning, wearing a shirt and tie and a tweed sport coat beneath his windbreaker, winced as the cold air hit him. With his thinning hair, wire-frame glasses, and pinched, disapproving manner, Drake conveyed a scholarly quality that in fact he cultivated. He did not want to be taken for what he was, a highly successful litigator who had retired to become the director of the National Environmental Resource Fund, a major American activist group. He had held the job at NERF for the last ten years.

Next, young Peter Evans bounced out of the car. Evans was the youngest of Morton's attorneys, and the one he liked best. Evans was twenty-eight and a junior associate of the Los Angeles firm of Hassle and Black. Now, even late at night, he remained cheerful and enthusiastic. He pulled on a Patagonia fleece and stuck his hands in his pockets, but otherwise gave no sign that the weather bothered him.

Morton had flown all of them in from Los Angeles on his Gulfstream G5 jet, arriving in Keflavík airport at nine yesterday morning. None of them had slept, but nobody was tired. Not even Morton, and he was sixty-five years old. He didn't feel the slightest sense of fatigue.

Just cold.

Morton zipped up his jacket and followed the graduate student down the rocky hill from the car. "The light at night gives you energy," the kid said. "Dr. Einarsson never sleeps more than four hours a night in the summer. None of us does."

"And where is Dr. Einarsson?" Morton asked.

"Down there." The kid pointed off to the left.

At first, Morton could see nothing at all. Finally he saw a red dot, and realized it was a vehicle. That was when he grasped the enormous size of the glacier.

Drake fell into step with Morton as they went down the hill. "George," he said, "you and Evans should feel free to go on a tour of the site, and let me talk to Per Einarsson alone."

"Why?"

"I expect Einarsson would be more comfortable if there weren't a lot of people standing around."

"But isn't the point that I'm the one who funds his research?"

"Of course," Drake said, "but I don't want to hammer that fact too hard. I don't want Per to feel compromised."

"I don't see how you can avoid it."

"I'll just point out the stakes," Drake said. "Help him to see the big picture."

"Frankly, I was looking forward to hearing this discussion," Morton said.

"I know," Drake said. "But it's delicate."

As they came closer to the glacier, Morton felt a distinct chill in the wind. The temperature dropped several degrees. They could see now the series of four large, tan tents arranged near the red Land Cruiser. From a distance, the tents had blended into the plain.

From one of the tents a very tall, blond man appeared. Per Einarsson threw up his hands and shouted, "Nicholas!"

"Per!" Drake raced forward.

Morton continued down the hill, feeling distinctly grouchy about being dismissed by Drake. Evans came up to walk alongside him. "I don't want to take any damn tour," Morton said.

"Oh, I don't know," Evans said, looking ahead. "It might be more interesting than we think." Coming out of one of the other tents were three young women in khakis, all blond and beautiful. They waved to the newcomers.

"Maybe you're right," Morton said.

Peter Evans knew that his client George Morton, despite his intense interest in all things environmental, had an even more intense interest in pretty women. And indeed, after a quick introduction to Einarsson, Morton happily allowed himself to be led away by Eva Jónsdóttir, who was tall and athletic, with short-cropped white blond hair and a radiant smile. She was Morton's type, Evans thought. She looked rather like Morton's beautiful assistant, Sarah Jones. He heard Morton say, "I had no idea so many women were interested in geology," and Morton and Eva drifted away, heading toward the glacier.

Evans knew he should accompany Morton. But perhaps Morton wanted to take this tour alone. And more important, Evans's firm also represented Nicholas Drake, and Evans had a nagging concern about what Drake was up to. Not that it was illegal or unethical, exactly. But Drake could be imperious, and what he was going to do might cause embarrassment later on. So for a moment Evans stood there, wondering which way to go, which man to follow.

It was Drake who made the decision for him, giving Evans a slight, dismissive wave of his hand as he disappeared into the big tent with Einarsson. Evans took the hint, and ambled off toward Morton and the girl. Eva was chattering on about how 12 percent of Iceland was covered in glaciers, and how some of the glaciers had active volcanoes poking out from the ice.

This particular glacier, she said, pointing upward, was of the type called a surge glacier, because it had a history of rapid advances and retreats. At the moment, she said, the glacier was pushing forward at the rate of one hundred meters a day—the length of a football field, every twenty-four hours. Sometimes, when the wind died, you could actually hear it grinding forward. This glacier had surged more than ten kilometers in the last few years.

Soon they were joined by Ásdis Sveinsdóttir, who could have been Eva's younger sister. She paid flattering attention to Evans, asking him how his trip over had been, how he liked Iceland, how long he was staying in the country. Eventually, she mentioned that she usually worked in the office at Reykjavík, and had only come out for the day. Evans realized then that she was here doing her job. The sponsors were visiting Einarsson, and Einarsson had arranged for the visit to be memorable.

Eva was explaining that although surge-type glaciers were very common—there were several hundred of them in Alaska—the mechanism of the surges was not known. Nor was the mechanism behind the periodic advances and retreats, which differed for each glacier. "There is still so much to study, to learn," she said, smiling at Morton.

That was when they heard shouts coming from the big tent, and considerable swearing. Evans excused himself, and headed back to the tent. Somewhat reluctantly, Morton trailed after him.

Per Einarsson was shaking with anger. He raised his fists. "I tell you, no!" he yelled, and pounded the table.

Standing opposite him, Drake was very red in the face, clenching his teeth. "Per," he said, "I am asking you to consider the realities."

"You are not!" Einarsson said, pounding the table again. "The reality is what you do not want me to publish!"

"Now, Per—"

"The reality" he said, "is that in Iceland the first half of the twentieth century was warmer than the second half, as in Greenland. The reality is that in Iceland, most glaciers lost mass after 1930 because summers warmed by 6 degrees Celsius, but since then the climate has become colder. The reality is that since 1970 these glaciers have been steadily advancing. They have regained half the ground that was lost earlier. Right now, eleven are surging. That is the reality, Nicholas! And I will not lie about it."

"No one has suggested you do," Drake said, lowering his voice and glancing at his newly arrived audience. "I am merely discussing how you word your paper, Per."

Einarsson raised a sheet of paper. "Yes, and you have suggested some wording—"

"Merely a suggestion—"

"That twists truth!"

"Per, with due respect, I feel you are exaggerating—"

"Am I?" Einarsson turned to the others and began to read. "This is what he wants me to say: 'The threat of global warming has melted glaciers throughout the world, and in Iceland as well. Many glaciers are shrinking dramatically, although paradoxically others are growing. However, in all cases recent extremes in climate variability seem to be the cause...blah...blah...blah... og svo framvegis.'" He threw the paperdown. "That is simply not true."

"It's just the opening paragraph. The rest of your paper will amplify."

"The opening paragraph is not true."

"Of course it is. It refers to 'extremes in climate variability.' No one can object to such vague wording."

"Recent extremes. But in Iceland these effects are not recent."

"Then take out 'recent.'"

"That is not adequate," Einarsson said, "because the implication of this paragraph is that we are observing the effects of global warming from greenhouse gases. Whereas in fact we are observing local climate patterns that are rather specific to Iceland and are unlikely to be related to any global pattern."

"And you can say so in your conclusion."

"But this opening paragraph will be a big joke among Arctic researchers. You think Motoyama or Sigurosson will not see through this paragraph? Or Hicks? Watanabe? Ísaksson? They will laugh and call me compromised. They will say I did it for grants."

"But there are other considerations," Drake said soothingly. "We must all be aware there are disinformation groups funded by industry—petroleum, automotive—who will seize on the report that some glaciers are growing, and use it to argue against global warming. That is what they always do. They snatch at anything to paint a false picture."

"How the information is used is not my concern. My concern is to report the truth as best I can."

"Very noble," Drake said. "Perhaps not so practical."

"I see. And you have brought the source of funding right here, in the form of Mr. Morton, so I do not miss the point?"

"No, no, Per," Drake said hastily. "Please, don't misunderstand—"

"I understand only too well. What is he doing here?" Einarsson was furious. "Mr. Morton? Do you approve of what I am being asked to do by Mr. Drake?"

It was at that point that Morton's cell phone rang, and with ill-concealed relief, he flipped it open. "Morton. Yes? Yes, John. Where are you? Vancouver? What time is it there?" He put his hand over the mouthpiece. "John Kim, in Vancouver. Scotiabank."

Evans nodded, though he had no idea who that was. Morton's financial operations were complex; he knew bankers all over the world. Morton turned and walked to the far side of the tent.

An awkward silence fell over the others as they waited. Einarsson stared at the floor, sucking in his breath, still furious. The blond women pretended to work, giving great attention to the papers they shuffled through. Drake stuck his hands in his pockets, looked at the roof of the tent.

Meanwhile, Morton was laughing. "Really? I hadn't heard that one," he said, chuckling. He glanced back at the others, and turned away again.

Drake said, "Look, Per, I feel we have gotten off on the wrong foot."

"Not at all," Einarsson said coldly. "We understand each other only too well. If you withdraw your support, you withdraw your support."

"Nobody is talking about withdrawing support . . ."

"Time will tell," he said.

And then Morton said, "What? They did what? Deposited to what? How much money are we—? Jesus Christ, John. This is unbelievable!" And still talking, he turned and walked out of the tent.

Evans hurried after him.

It was brighter, the sun now higher in the sky, trying to break through low clouds. Morton was scrambling up the slope, still talking on the phone. He was shouting, but his words were lost in the wind as Evans followed him.

They came to the Land Cruiser. Morton ducked down, using it as a shield against the wind. "Christ, John, do I have legal liability there? I mean—no, I didn't know a thing about it. What was the organization? Friends of the Planet Fund?"

Morton looked questioningly at Evans. Evans shook his head. He'd never heard of Friends of the Planet. And he knew most of the environmental organizations.

"Based where?" Morton was saying. "San Jose? California? Oh. Jesus. What the hell is based in Costa Rica?" He cupped his hand over the phone. "Friends of the Planet Fund, San José, Costa Rica."

Evans shook his head.

"I never heard of them," Morton said, "and neither has my lawyer. And I don't remember—no, Ed, if it was a quarter of a million dollars, I'd remember. The check was issued where? I see. And my name was where? I see. Okay, thanks. Yeah. I will. Bye." He flipped the phone shut.

He turned to Evans.

"Peter," he said. "Get a pad and make notes."

Morton spoke quickly. Evans scribbled, trying to keep up. It was a complicated story that he took down as best he could.

John Kim, the manager of Scotiabank, Vancouver, had been called by a customer named Nat Damon, a local marine operator. Damon had deposited a check from a company called Seismic Services, in Calgary, and the check had bounced. It was for $300,000. Damon was nervous about whoever had written the check, and asked Kim to look into it.

John Kim could not legally make inquiries in the U.S., but the issuing bank was in Calgary, and he had a friend who worked there. He learned that Seismic Services was an account with a postal box for an address. The account was modestly active, receiving deposits every few weeks from only one source: The Friends of the Planet Foundation, based in San José, Costa Rica.

Kim placed a call down there. Then, about that time, it came up on his screen that the check had cleared. Kim called Damon and asked him if he wanted to drop the inquiry. Damon said no, check it out.

Kim had a brief conversation with Miguel Chavez at the Banco Credito Agricola in San José. Chavez said he had gotten an electronic deposit from the Moriah Wind Power Associates via Ansbach (Cayman) Ltd., a private bank on Grand Cayman island. That was all he knew.

Chavez called Kim back ten minutes later to say he had made inquiries at Ansbach and had obtained a record of a wire transfer that was paid into the Moriah account by the International Wilderness Preservation Society three days before that. And the IWPS transfer noted in the comment field, "G. Morton Research Fund."

John Kim called his Vancouver client, Nat Damon, to ask what the check was for. Damon said it was for the lease of a small two-man research submarine.

Kim thought that was pretty interesting, so he telephoned his friend George Morton to kid him a bit, and ask why he was leasing a submarine. And to his surprise, Morton knew absolutely nothing about it.

Evans finished taking down notes on the pad. He said, "This is what some bank manager in Vancouver told you?"

"Yes. A good friend of mine. Why are you looking at me that way?"

"Because it's a lot of information," Evans said. He didn't know the banking rules in Canada, to say nothing of Costa

Rica, but he knew it was unlikely that any banks would freely exchange information in the way Morton had described. If the Vancouver manager's story was true, there was more to it that he wasn't telling. Evans made a note to check into it. "And do you know the International Wilderness Preservation Society, which has your check for a quarter of a million dollars?"

Morton shook his head. "Never heard of them."

"So you never gave them two hundred and fifty thousand dollars?"

Morton shook his head. "I'll tell you what I did do, in the last week," he said. "I gave two hundred and fifty grand to Nicholas Drake to cover a monthly operating shortfall. He told me he had some problem about a big contributor from Seattle not coming through for a week. Drake's asked me to help him out before like that, once or twice."

"You think that money ended up in Vancouver?"

Morton nodded.

"You better ask Drake about it," Evans said.

"I have no idea at all," Drake said, looking mystified. "Costa Rica? International Wilderness Preservation? My goodness, I can't imagine."

Evans said, "You know the International Wilderness Preservation Society?"

"Very well," Drake said. "They're excellent. We've worked closely with them on any number of projects around the world— the Everglades, Tiger Tops in Nepal, the Lake Toba preserve in Sumatra. The only thing I can think is that somehow George's check was mistakenly deposited in the wrong account. Or . . . I just don't know. I have to call the office. But it's late in California. It'll have to wait until morning."

Morton was staring at Drake, not speaking.

"George," Drake said, turning to him. "I'm sure this must make you feel very strange. Even if it's an honest mistake—as I am almost certain it is—it's still a lot of money to be mishandled. I feel terrible. But mistakes happen, especially if you use a lot of unpaid volunteers, as we do. But you and I have been friends for a long time. I want you to know that I will get to the bottom of this. And of course I will see that the money is recovered at once. You have my word, George."

"Thank you," Morton said.

They all climbed into the Land Cruiser.

The vehicle bounced over the barren plain. "Damn, those Icelanders are stubborn," Drake said, staring out the window. "They may be the most stubborn researchers in the world."

"He never saw your point?" Evans said.

"No," Drake said, "I couldn't make him understand. Scientists can't adopt that lofty attitude anymore. They can't say, 'I do the research, and I don't care how it is used.' That's out of date. It's irresponsible. Even in a seemingly obscure field like glacier geology. Because, like it or not, we're in the middle of a war—a global war of information versus disinformation. The war is fought on many battlegrounds. Newspaper op-eds. Television reports. Scientific journals. Websites, conferences, classrooms— and courtrooms, too, if it comes to that." Drake shook his head. "We have truth on our side, but we're outnumbered and outfunded. Today, the environmental movement is David battling Goliath. And Goliath is Aventis and Alcatel, Humana and GE, BP and Bayer, Shell and Glaxo-Wellcome—huge, global, corporate. These people are the implacable enemies of our planet, and Per

Einarsson, out there on his glacier, is irresponsible to pretend it isn't happening."

Sitting beside Drake, Peter Evans nodded sympathetically, though in fact he took everything Drake was saying with a large grain of salt. The head of NERF was famously melodramatic. And Drake was pointedly ignoring the fact that several of the corporations he had named made substantial contributions to NERF every year, and three executives from those companies actually sat on Drake's board of advisors. That was true of many environmental organizations these days, although the reasons behind corporate involvement were much debated.

"Well," Morton said, "maybe Per will reconsider later on."

"I doubt it," Drake said gloomily. "He was angry. We've lost this battle, I'm sorry to say. But we do what we always do. Soldier on. Fight the good fight."

It was silent in the car for a while.

"The girls were damn good-looking," Morton said. "Weren't they, Peter?"

"Yes," Evans said. "They were."

Evans knew that Morton was trying to lighten the mood in the car. But Drake would have none of it. The head of NERF stared morosely at the barren landscape, and shook his head mournfully at the snow-covered mountains in the distance.

Evans had traveled many times with Drake and Morton in the last couple of years. Usually, Morton could cheer everybody around him, even Drake, who was glum and fretful.

But lately Drake had become even more pessimistic than usual. Evans had first noticed it a few weeks ago, and had wondered at the time if there was illness in the family, or something else that was bothering him. But it seemed there was nothing amiss. At least, nothing that anyone would talk about. NERF was a beehive of activity; they had moved into a wonderful new building in Beverly Hills; fund-raising was at an all-time high; they were planning spectacular new events and conferences, including the Abrupt Climate Change Conference that would begin in two months. Yet despite these successes—or because of them?—Drake seemed more miserable than ever.

Morton noticed it, too, but he shrugged it off. "He's a lawyer," he said. "What do you expect? Forget about it."

By the time they reached Reykjavík, the sunny day had turned wet and chilly. It was sleeting at Keflavík airport, obliging them to wait while the wings of the white Gulfstream jet were de-iced. Evans slipped away to a corner of the hangar and, since it was still the middle of the night in the US, placed a call to a friend in banking in Hong Kong. He asked about the Vancouver story.

"Absolutely impossible," was the immediate answer. "No bank would divulge such information, even to another bank. There's an STR in the chain somewhere."

"An STR?"

"Suspicious transfer report. If it looks like money for drug trafficking or terrorism, the account gets tagged. And from then on, it's tracked. There are ways to track electronic transfers, even with strong encryption. But none of that tracking is ever going to wind up on the desk of a bank manager."

"No?"

"Not a chance. You'd need international law-enforcement credentials to see that tracking report."

"So this bank manager didn't do all this himself?"

"I doubt it. There is somebody else involved in this story. A policeman of some kind. Somebody you're not being told about."

"Like a customs guy, or Interpol?"

"Or something."

"Why would my client be contacted at all?"

"I don't know. But it's not an accident. Does your client have any radical tendencies?"

Thinking of Morton, Evans wanted to laugh. "Absolutely not."

"You quite sure, Peter?"

"Well, yes . . ."

"Because sometimes these wealthy donors amuse themselves, or justify themselves, by supporting terrorist groups. That's what happened with the IRA. Rich Americans in Boston supported them for decades. But times have changed. No one is amused any longer. Your client should be careful. And if you're his attorney, you should be careful, too. Hate to visit you in prison, Peter."

And he hung up.

*Of all Al Gore's contributions to the global warming debate, none is as important as* An Inconvenient Truth, *the 2006 documentary that was based on the slide shows Gore had showed to audiences around the world in the years since his defeat in the 2000 presidential election. The movie, which won an Academy Award (and propelled Gore to his Nobel Prize) is essentially a PowerPoint presentation brought to life; it showed audiences that the science was much more compelling than they'd previously understood, and helped make climate change an important issue once more. It was, of course, harshly attacked by climate skeptics, but with a few small exceptions scientists vouched for its accuracy and moderation. The only real criticisms leveled at the film were for proposing too few, and too modest, changes: many viewers changed their lightbulbs and then were left wondering what more to do.*

# from *An Inconvenient Truth*

Al Gore
2006

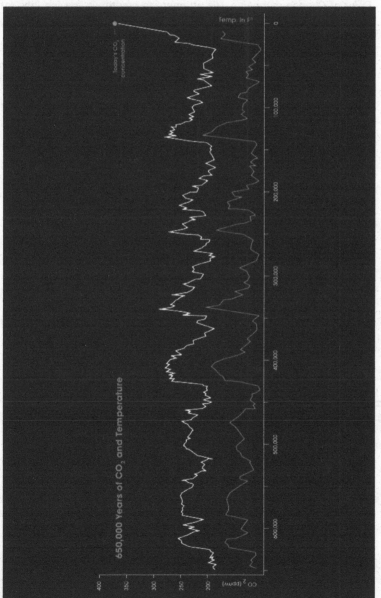

650,000 Years of CO₂ and Temperature

Temp. in F°

Today's CO₂ concentration

CO₂ (ppmv)

*Van Jones is a remarkable activist. A radical in his early years (the verdict in the Rodney King police brutality case, he said, briefly turned him into a "communist"), he matured into a leading human rights and social justice activist, organizing the Ella Baker Center in Oakland and the nationwide Color of Change campaign. His attempts to create well-paying jobs for inner-city residents led him to found the organization Green For All, which works toward a new green economy that leaves no one behind. His book The Green Collar Economy reached the twelfth spot on the New York Times bestseller list in 2008, and Jones was named to a White House job soon after Barack Obama's inauguration. But the right wing focused on his past rhetoric, and, attacked with a harrowing smear campaign on conservative radio and TV, he resigned his post after only six months on the job. Nonetheless, he continues to be one of the most visible leaders of the green progressive movement.*

# from *The Green Collar Economy: How One Solution Can Fix Our Two Biggest Problems*

Van Jones with Ariane Conrad
*2008*

*Fewer "Targets," More Partners*

The other problem with the standard "organizer's" formulation is the constant seeking-out of "targets" to pressure. Again, sometimes there is a person in a position of authority who is so obstinate, so biased, and so recalcitrant that one has no choice but to declare him or her an opponent. Also, there are institutions that have acted in bad faith for so long that trust is almost impossible to regenerate. Under those circumstances, the advocates of a righteous cause do need a battle-hardened cadre of well-trained organizers who know how to twist pinkies and otherwise force an adversary into submission. The protectors of the status quo use power tactics all the time. The champions of a better tomorrow should not unilaterally disarm.

Yet this approach can be overdone, overplayed, and overused. Too often activists just assume that any change worth making will always require a big battle with someone. They start preparing for Armageddon every time any issue comes up—even before they have taken the first steps to resolve it using less confrontational means.

Some organizations are like countries run by generals who have an army but no diplomatic corps. Therefore, they spend all their time drilling their troops and scanning the horizon, hoping for an opportunity to declare war on someone or something. Again, sometimes this is justifiable.

However, it comes down to a question of balance. When all you have is a hammer, everything looks like a nail. If all you have is "direct-action organizing," everyone with power looks like a target. It does take real skill, talent, and training to identify targets, challenge them, and get them to do what you want. The time has come for social change and environmental organizations to add another set of skills: the ability to turn would-be "targets" into real, long-term partners for change. And that can be a tougher challenge.

For one thing, it requires that organizers move beyond assumptions, stereotypes, and past hurts. It requires that organizers (and those they organize) invest time in relationship-building and trust-building across lines of race, class, and authority—trying to surface points of shared interest and concern. Not every grassroots group has the time, capacity, or organizational strength to function in this way. Sometimes it is easier for marginalized activists to just call a press conference and start painting protest signs. I understand that. Yet over the long term, the accumulated results often are not worth all the expended effort.

Here's the truth. If you rush into a situation looking for enemies, you will always find plenty. At the same time, if you go into a situation trying to find friends and allies, you will almost always find at least one. Sometimes they are in surprisingly powerful places—like behind the receptionist's desk in your opponent's front office.

In this age, our main job is to seek out friends wherever we can, not just to defeat enemies. What if, rather than mainly looking for opponents to punish, those of us who are committed to social change spent our time seeking out potential allies to encourage, befriend, and reward. After all, for every scofflaw polluter, there may be dozens of local businesses out there trying to do the right thing ecologically—but getting little support or recognition. What if environmentalists did more to partner with them, celebrate them, and help them? For every racist employer or bigoted beat cop, there are tens of thousands of white people who absolutely abhor racism. And yet civil rights activists like myself rarely ask them to do anything—except to feel guilty. Why not focus on finding better ways to access their time and talent for the good of all?

Our cause needs fewer enemies and more friends. To get through the coming crises, we are going to need each other. Let's start laying the groundwork now, so that—later on—we will be more likely to turn to each other, not on each other.

## Less "Accusation," More Confession

None of this is to say that we won't have to confront and defeat real, implacable, and unyielding enemies on our journey. We will. If we could achieve eco-equity only by defeating external enemies, we would be walking an easier path than the one we're on.

That's because some of the enemies we need to defeat are inside us. We ourselves are a part of the problem. Every day, almost all of us are working and consuming in the pollution-based economy. We are participating in an economy that lacks equity, and yet we each have an understandable aversion to giving up our own money or status. We are trying to change the status quo. But we all have a stake in it too. We all rely upon it to live and survive. And so, every day, we end up feeding the very monster we are fighting.

This is a humbling fact. If we are honest, even those of us who desperately want change must admit that we are not just battling the polluter without; we are also battling the polluter within. We can say this not just in the obvious "material" sense. This is not just about how many times we fail to recycle, bicycle, or bring our cloth bag to the grocery store. We all have inner demons that pollute our minds and hearts—that cloud our thoughts and distort our actions.

As we try to work with others, our egos often get in the way. Tempers flare. Suddenly we find ourselves not just battling the warmonger without, but that white-hot warmonger within. Later on, resentment creeps in over some perceived sleight [sic] and indignity. Eventually, we find ourselves battling not just the punitive, unforgiving jailers on the outside. We end up battling the punitive jailer within, the hurt and angry part of ourselves who can't forgive our coworkers and allies for shortcomings and disappointments. These are the hidden struggles that define our days. Cumulatively, these inner tumults determine and limit the impact of our work itself, but nobody talks about them much.

Instead, we engage in the old politics, naming, blaming, and shaming somebody else while concealing our own faults, flaws, and hypocrisies. However, the cause of pursuing eco-equity does not easily lend itself to that approach. The change we are seeking is too monumental, and our own capacities are too modest.

We would be better off confessing our own weaknesses, our fears, our needs. Doing so will let others see the gaps more quickly, find their rightful places around the growing circle—and come to the campfire with fewer pretenses themselves. If we confess our own struggles to realign our own lives and change our own behavior, we may seem less alien to those we are trying to convince.

Also, the change we seek is so complex that no one person can understand everything that must be done. In that regard, we are all equally ignorant about how to get where we are going. This weakness actually is our strength. If we confess our own uncertainty, we are much more likely to listen attentively to others—and pull others into speaking more honestly and fully. As we move forward, our motto should be: accuse less, confess more.

### Less "Cheap Patriotism," More Deep Patriotism

We have gone through a period during which people waving American flags have done great damage to the country, to the people of Iraq, to America's prestige in the world, to the national treasury, to the U.S. Constitution, and to the international rule of law. While force-feeding the country a brand of cheap and mindless patriotism, the "leaders" waving the biggest flags have steered the nation into a ditch. People of conscience should embrace Old Glory—and use the flag to help guide the public back in the direction of sanity.

One begins to fear that this accident was not very accidental. After all, GOP anti-tax operative Grover Norquist had declared openly: "I don't want to abolish government. I simply want to reduce it to the size where I can drag it into the bathroom and drown it in the bathtub."[1] That is not a patriotic statement.

We have an obligation to tell the ultraconservatives who are so rabidly antigovernment: "If you don't love this government, then let it go and hand it over to people who do." Those who would hijack the government and crash it with deficits pose a bigger threat than the terrorists.

And while we are at it, we could make do with a lot less knee-jerk antipatriotism from the left. I know it is hard to make peace with the country's original sins of stolen land and stolen labor. It is hard to forgive its repeated entanglement in unjust wars, up to the present moment. However, the far left's strategy of trying to fix the country by putting it down all the time has been an utter failure.

To paraphrase scholar Cornel West, you can't save a country you don't serve, and you can't lead a country you don't love. And there is much to love in this country. After all, we are talking about the nation that gave the world basketball, iPods, and Beyoncé Knowles. (If those three won't get you up stomping and cheering for the red, white, and blue, I don't know what will.)

The United States has the power to be a huge obstacle to planetary survival—or giant springboard to planetary salvation. A better America is the best gift that we can offer the world. Yet caring Americans will never give the world that gift if they are holding their noses and handling the flag like a used tissue.

If we do our work right, the United States will lead the world, again, someday. This next time—not in war. Not in per capita greenhouse-gas emissions. Not in incarceration rates. The United States will lead the world in green economic development, in world-saving technologies, in human rights. We will lead by showing a multiracial, multifaith, rainbow-colored planet how our multiracial, multifaith, rainbow-colored country pulled together to solve tough problems. The United States will go from being the world leader in ecological pollution to the world leader in ecological solutions.

Bruce Springsteen put it best in 2004 when he said: "America is not always right. That's a fairy tale for children.... But one thing America should always be is true. And it's in seeking her truth, both the good and the bad, that we find a deeper patriotism, that we find a more authentic experience as citizens, that we find the power that is embedded only in truth to change our world for the better."[2]

It's time for the deeply patriotic to take back the flag from the cheaply patriotic, because, despite the pain of old crimes and recent disappointments, some of us still believe in America. Some of us still believe in "a more perfect union"—and in making it more perfect every day. Some of us still believe in "America the beautiful"—and in defending its beauty from the clear-cutters and despoilers. Some of us still believe in "one nation, indivisible"—and in opposing those who profit by keeping us needlessly divided. Some of us still believe in "liberty and justice for all," and we won't stop until that classroom pledge is honored from shore to shore.

Some of us still believe in America—and in all of those things we learned about it as children. Of course, we know now that America is not the place we live, but a destination to which we all are headed. So we keep faith on the journey. No, some of us haven't given up on Dr. King's dream. There are those of us who yet believe we are going win.

And when we do, we'll be doing more than just "taking America back." We will be taking America—forward.

## The Future Is Now

We have explored the principles and the politics that could revive the economy on a more inclusive and ecologically responsible basis. Luckily for us, leaders in far-flung and unlikely places are already moving ahead and creating this future. They are not waiting for federal action to give them the green light to start creating the new economy.

We will examine courageous pioneers who are already helping ordinary people blaze green pathways to prosperity. I hope that someday the vast majority of U.S. workers will have jobs in the kinds of innovative enterprises and programs that we explore below. After all, every day, about 145 million people go to work in the United States.[3] Imagine if those jobs—plus new ones created for people who are currently unemployed—were largely working in fields and professions that uplift human dignity and honor the Earth.

Some are already doing it today. Here are their stories, organized by five major subsystems of sustainability: energy, food, waste, water, and transportation.

The transition from our reliance on fossil fuels to clean and renewable energy is the linchpin of the green economy. If done correctly, it will bring our carbon emissions down to a manageable level. It will free us from foreign oil and its national security risks. It will halt the skyrocketing rates of pollution-based illnesses, and it will revitalize our economy and create millions of green-collar jobs.[4] Here are projects that are simultaneously producing clean energy and creating career paths for the unemployed.

*Energy Efficiency*

The cleanest energy is the energy that we never have to use— because we were wise enough to conserve it. We waste a lot—in heating, cooling, and lighting our drafty, poorly designed homes and offices. Not to mention driving our cars, with their unimpressive number of miles per gallon. The energy we don't use is cheap, silent, clean. It's measured in "negawatts," instead of megawatts. Improvements in the energy efficiency of buildings (such as weather stripping, replacing fixtures, and insulating hot-water heaters) can simultaneously save property-owners money, reduce demand for fossil-fuel-generated electricity, and provide both skilled and unskilled jobs for local workers.[5]

In Los Angeles, the community-based group Strategic Concepts in Organizing and Policy Education (SCOPE) convened the local Apollo Alliance. Campaign Coordinator at SCOPE, Elsa Barboza, says the local Alliance's first step was "To collect signatures from black, Latino, Asian, and Anglo working-class families throughout Los Angeles's inner-city neighborhoods for a petition to create a sustainable, equitable, and clean energy economy that will bring quality jobs to their communities, create a healthier and safer environment, and promote community-based land use planning and economic development."[6]

One of the people out knocking on doors with the petition was Oreatha Ensley, a lifelong civil rights activist, a mother and grandmother, a former teacher, and an L.A. resident for nearly forty years. She says: "I expected some folks to tell me that jobs are number one and cleaning our environment is just a nice wish. Instead, they told me that it's about time we reinvest in our community, because we are slipping away further into poverty and getting sicker because of it."[7]

It wasn't just the poor communities who were in favor of the Alliance's objectives. "Mayor Villaraigosa, a liberal-leaning city council, and forward-thinking commissioners have articulated a bold vision to make Los Angeles a national leader in the transition to a sustainable, equitable, clean energy economy," Elsa notes. Indeed, the City of Los Angeles has already committed to one of the local Apollo Alliance's proposals: implementation of a pilot program to retrofit one hundred city-owned buildings with energy and water-conservation technologies.[8]

Nationwide, buildings are responsible for thirty-six percent of our energy use, thirty percent of our greenhouse-gas emissions, and thirty percent of our waste production. Once complete, retrofits in L.A. are slated to save the city up to $10 million per year in utility costs.[9]

The City of Los Angeles owns and operates more than eleven hundred buildings, many in deteriorating condition, that cover over a million square feet. The work includes audits; energy-efficiency improvements (e.g., sealing around or replacing doors and windows); lighting upgrades (replacing bulbs, installing sensors, and maximizing daylight); water-conservation improvements (fixing leaks, replacing urinals and toilets); heating- and cooling-system updates. And cool or green-roofing installation audits—the first step in the process—are under way in L.A. as I write this.[10]

In the longer term, workers from the retrofit jobs can be transitioned into maintenance and construction jobs in both public and private sectors. According to the California Employment Development Department, employment opportunities in construction in the L.A. area are projected to increase by thirty percent between 2002 and 2012. The industry's career ladders allow workers earning entry-level wages of $9 to $18 per hour to become advanced skilled workers such as plumbers, sheet-metal workers, and electricians, earning $15 to as much as $50 per hour.[11]

Meanwhile, in the old industrial city of Milwaukee, Wisconsin, an organization called the Center on Wisconsin Strategy (COWS) is exploring how workers in the Rust Belt can move to the center of the clean-energy economy. COWS has cooked up a brilliant scheme to retrofit all of Milwaukee's buildings—and to create a slew of green-collar jobs in the process. To make this work, a new building-efficiency service—Milwaukee Energy Efficiency (Me2)—is being created to offer all of that city's residents the opportunity to buy and install cost-effective energy-efficiency measures in their homes and businesses.

Here's the beautiful part: there will be no up-front payment, no new debt obligation. Customers will have full assurance that their utility costs will be lower, and they will make monthly payments only for as long as they remain at the location and the measures continue to work. Outside of *The Godfather*, that's about as close as you can get in this life to "an offer they can't refuse."

This is how it will work. Owners or renters sign up to have their place retrofitted to save energy costs; a qualified person shows up and does the work; and then the renters or owners pay off the cost of the retrofit a little bit at a time, over the course of years, as a part of their (now radically reduced) electricity or property-services bill. That's it. Everybody wins—including the Earth.

In the policy section that follows, we will go into more detail about how this program works. For now, suffice it to say that the Me2 program will be great for those community residents seeking jobs. COWS and the University of Florida estimate that every $1 million spent on the effort will generate about ten job-years in installation and construction activities and another three job-years in upstream manufacturing for needed parts. For a roughly

$500 million project, that's a lot of job-years: about sixty-five hundred. Residents can start with less-skilled work—like blowing insulation or wrapping pipes—and move up to the more advanced work in plumbing, wiring, and installing new heating and ventilation systems.[12]

Retrofitting work will provide good jobs—family-supporting gigs, with solid opportunities for advancement—that cannot be outsourced. They will feed into the exploding green building industry, and that will simultaneously reduce our emissions and our reliance on foreign oil. Similar energy-efficiency programs should be implemented across the country.

*Billy Parish grew up in New York City, but a semester at the Mountain School in Vermont helped turn him into an environmentalist. He dropped out of Yale University in 2003 to help start the Energy Action Coalition (EAC), the main vehicle for youth organizing on climate issues in the United States. A tireless organizer with a special gift for mediating diverse viewpoints, Parish emerged as a generational spokesman on the issue. Arguably, the EAC has helped spur the biggest organizing drive among college students since the height of the anti-apartheid movement. Enormous biennial Power Shift gatherings have helped the movement coalesce, and many of Parish's former colleagues now run important parts of the green movement.*

# Climate Generation:
# The Evolution of the Energy Action
# Coalition's Strategy

Billy Parish
2000

This Climate Generation series is well timed. Like every other national group I know working on the transition to a clean energy economy, the Energy Action Coalition is going through a strategic planning/soul-searching process to figure out just what the &$*$ to do next. The outline of the broader movement's situation is pretty well understood, but few really good ideas about how to solve it have surfaced. Despite some meaningful accomplishments in 2009—mainly through the Recovery Act and executive actions—the big goals of passing strong federal legislation before Copenhagen and securing a binding international treaty have not been achieved.

The chasm between what's needed and what is being discussed in Washington grows ever wider, and the "pragmatists" inside and outside the Beltway can barely hear each other anymore. Every week comes with more dire scientific predictions and, newly, worse polling numbers on public understanding of the impacts and support for action. Major Democratic losses predicted[1] for the 2010 midterm elections confirm the widespread feeling that our golden opportunity for change is slipping away.

We're still just not powerful enough as a movement to make the changes we so desperately need. As Jamie [Henn]'s great post[2] yesterday clearly laid out, we need to be thinking about strategies that go big. To complement Jamie's history, I want to sketch out my understanding of how the Energy Action Coalition's strategy has evolved over the past six years with the hope that a better understanding of our strategic history can inform the decisions we make moving forward.

Three quick notes before I do: 1) I believe the coalition's collaborative planning processes—and culture of fun, diversity, aspirational thinking, solidarity, and action—have been a large part of how we've developed smart strategy, but the focus here is on the results and what we actually did, not how we came to the decisions; 2) I'd also recommend a look at Sara Robinson's account[3] of the progression of social change, which provides a broader context in which to situate these decisions; and 3) This is *my* interpretation of events, biased as it surely is.

*Phase 1: Finding Ourselves (November 2003–August 2005)*

The story begins with Campus Clean Energy Day (November 13, 2003), the first distributed day of action of this fledgling movement with sixty-five schools participating nationwide. It was a zeitgeisty kind of thing—all across the country, filling the void of leadership from politicians and corporations, student groups were beginning to run campaigns to get their schools to purchase clean energy. Student networks like SEAC, EnviroCitizen, Greenpeace, and the Climate Campaign built on this interest and planned a day to demonstrate the growing interest in student activism on climate solutions. Our second day of action, Fossil Fools Day on April 1, 2004, had exactly twice as many events.

The relationships between the full-time organizers and national student leaders grew stronger and a founding meeting was planned for the first week of June in Washington. Leaders from the seventeen organizations present at the meeting developed a founding mission statement: "To unite a diversity of organizations that will support and strengthen the student and youth clean energy movement in Canada and the U.S. Together we will leverage our collective power and create change for a clean, efficient, just and renewable energy future. We will accomplish this by focusing on four strategic areas: campuses, communities, corporate practices, and politics."

Faced with the upcoming '04 presidential elections, the coalition's first campaign centered around a Declaration of Independence from Dirty Energy pledge and culminated with 280 youth-organized events on October 19, 2004, for Energy Independence Day. We gathered nearly 30,000 pledge signatures, organized hundreds of events to educate young people on the failures of Bush's energy and climate policies and weakly engaged in some voter registration and get-out-the-vote operations. We had basically no money, no way to reach out to our base with a clear message and campaign; there were only a handful of full-time paid organizers working on the issue; and we were still developing the structures and processes to make collective decisions.

After the shock and trauma of that election night faded, we reconvened in Washington in January with eighty leaders representing a much broader and more diverse cross section of our movement to strategize together about what to do next. Inspired by the Georgian "Orange Revolution" going on as we met and seeking a way to build our power from the bottom up, the idea was first thrown out to develop a unified campaign around campus sustainability, modeled in some ways on what the Sierra Youth Coalition had created in Canada. Over the next eight months, in addition to several days of action and a waste-vege-table-powered bus tour gathering another 30,000 signatures urging the Big Three auto makers to build cleaner cars, the Campus Climate Challenge campaign was developed and initial fundraising had begun.

This period was generally characterized by reactive and opportunistic strategy development, a heavy focus on one-off days of action (see Josh Lynch's EAC Day of Action history from 11/03 to 2/07 for more)[4] and a resulting weak movement narrative. We were just beginning to find each other, just beginning to trust each other.

## Phase 2: Building Power (September 2005–October 2007)

The Campus Climate Challenge was conceived of as a three-year campaign to make our schools models of the sustainable future we wanted for our country. The idea was to support local campaigns that could win important victories; recruit, train, and empower large numbers of youth organizers and consolidate our base into a single broad campaign that could be mapped on a website. It was the heart of the Bush years and we were building our base. Coalition partners had the flexibility to work with local groups to then leverage their campus campaigns and victories to stop dirty energy projects from being built, influence local and state policy, or run corporate campaigns. Days of Action throughout this period built on strategic themes within the campaign and demonstrated a rapidly growing number of groups across the country working on these issues.

Key to developing a campaign that could work for our then-thirty-odd coalition partners was creating a joint fundraising and budgeting process that all partners could participate in.

We soft-launched the campaign in September '05 and continued to raise money, recruit campuses, and create the materials
throughout that year. By the summer of '06, we had raised $3
million and began hiring what soon became seventy-five FTE
staff across coalition partner organizations and the central staff
to run the campaign. Two indigenous organizations created the
Tribal Campus Climate Challenge, another ran the campaign on
Historically Black Colleges and Universities, another developed
an endowment strategy, another organized schools into state networks to impact state policy, and on and on.

Nearly 1,000 youth groups around the country ran the campaign, and large state and regional trainings provided concrete
organizing skills, built state networks, and allowed the youth
leaders to share ideas and plan larger campaigns together. The
victories started to come in—a ten percent clean energy purchase
here, a commitment to Kyoto-level carbon reductions at another
college there —but they were too few and too little. At a coalition
strategy meeting during the summer of '06 we developed an idea
for a college president's version of the Challenge, which a partner organization and two allied groups turned into the American
College and University Presidents' Climate Commitment.[5] The
combination of bottom-up student organizing and top-down president leadership was a game changer, and as of this writing, 666
colleges have committed to becoming climate neutral. Spooky
number, but amazing results.

Young people were also increasingly frustrated with the lack
of leadership from just about anywhere else and began to escalate tactics—a sit-in at the Penn State president's office that was
replicated elsewhere, actions targeting existing and proposed
coal plants, a fast in front of the White House, and the Climate
Summer marches in Vermont and New Hampshire. The Step It

Up 2007 campaign,[6] led by an amazing group of Middlebury students and author Bill McKibben, took the basic concept of a day of action to a new level. By focusing on a single political task and going intergenerational, they were able to mobilize 1,400 events around the country—more than double the size of any previous Energy Action Coalition–supported day of action.

This period in the coalition's growth was the beginning of a sustained, coordinated strategy that leveraged the strengths of almost all our coalition partners. But "The Challenge" was at core a campus campaign, and didn't reach the two-thirds of youth eighteen to twenty-four who weren't in school. It was also highly decentralized, making it hard to demonstrate our collective power—to our targets, and even ourselves.

*Phase 3: Pushing the National Debate (November 2007–present)*

The first Power Shift in early November 2007 was our entrance onto the national stage. Six thousand young people in D.C. was at that point the largest climate advocacy gathering in the country's history. We developed a national platform—(1) Create Five Million New Green Jobs, (2) Cut Carbon Eighty Percent by 2050, and (3) A Moratorium on New Coal Plants and an End to Fossil Fuel Subsidies—and swamped Capitol Hill. All three goals, championed by the youth movement while most others thought they were impossibly radical, have become mainstream, and are in fact endorsed by our current president.

Power Shift '07 was the beginning of the coalition's focus on the '08 elections with the Power Vote campaign. The most disciplined and coordinated coalition campaign yet generated huge results: 350,000 signatures on a Power Vote pledge, hundreds of events and media articles, and a strong presence at the conventions and presidential debates. The coalition began to develop strong online organizing skills and became major power player in Washington, D.C.

To call on the leaders we had helped elect to make good on their promises and demonstrate our continued resolve and power, the coalition recruited nearly 13,000 people to Washington for Power Shift '09 in early March. The huge lobby day and largest-ever climate civil disobedience action—the Capitol Climate Action, organized by allied groups in partnership with the coalition—represented a new level for the movement as a whole and sent an important early signal to the new administration and Congress.

After Power Shift, the coalition's focus on federal policy has continued, with mini–Power Shifts in key Senate states, calling and letter-writing campaigns, and creative local actions. The "It's Game Time, Obama" campaign, one of the only large-scale public efforts to call out the president's lack of leadership on the issue, has succeeded on two of its three asks of him: (1) attending Copenhagen and (2) organizing a meeting between top youth leaders and five of his cabinet secretaries, but not (3) making an address to the nation specifically on climate and energy.

And so here we are. Another critical moment stands before us where the broader movement needs the vision and leadership of young people. What does the end of Phase 3 look like? Or Phase 4 or 5? Who will step up to lead us to a set of strategies that allow us to go big, to multiply our power, and win the victories our civilization needs to survive?

*A veteran local climate organizer, Mike Tidwell founded the Chesapeake Climate Action Network, which has won many legislative battles in Maryland and Virginia. Though he himself lives an exemplary environmental life (heating his home with the waste from corn farms, for instance), Tidwell wrote this powerful essay rejecting much of the green consumerism that has sprung up in the wake of* An Inconvenient Truth. *Or not so much rejecting it as arguing that it should come second to building a movement capable of forcing real change.*

# To Really Save the Planet, Stop Going Green

Mike Tidwell
*2009*

As President Obama heads to Copenhagen next week for global warming talks, there's one simple step Americans back home can take to help out: stop "going green." Just stop it. No more compact fluorescent light bulbs. No more green wedding planning. No more organic toothpicks for holiday hors d'oeuvres.

December should be national Green-Free Month. Instead of continuing our faddish and counterproductive emphasis on small, voluntary actions, we should follow the example of Americans during past moral crises and work toward large-scale change. The country's last real moral and social revolution was set in motion by the civil rights movement. And in the 1960s, civil rights activists didn't ask bigoted Southern governors and sheriffs to consider "Ten Ways to Go Integrated" at their convenience.

Green gestures we have in abundance in America. Green political action, not so much. And the gestures ("Look honey, another *Vanity Fair* 'Green Issue'!") lure us into believing that broad change is happening when the data shows that it isn't. Despite all our talk about washing clothes in cold water, we aren't making much of a difference.

For eight years, George W. Bush promoted voluntary action as the nation's primary response to global warming—and for eight years, aggregate greenhouse gas emissions remained unchanged. Even today, only ten percent of our household light bulbs are compact fluorescents. Hybrids account for only 2.5 percent of U.S. auto sales. One can almost imagine the big energy companies secretly applauding each time we distract ourselves from the big picture with a hectoring list of "Five Easy Ways to Green Your Office."

As America joins the rest of the world in finally fighting global warming, we need to bring our battle plan up to scale. If you believe that astronauts have been to the moon and that the world is not flat, then you probably believe the satellite photos showing the Greenland ice sheet in full-on meltdown. Much of Manhattan and the Eastern Shore of Maryland may join the Atlantic Ocean in our lifetimes. Entire Pacific island nations will disappear. Hurricanes will bring untold destruction. Rising sea levels and crippling droughts will decimate crops and cause widespread famine. People will go hungry, and people will die.

Morally, this is sort of a big deal. It would be wrong to let all this happen when we have the power to prevent the worst of it by adopting clean-energy policies.

But how do we do that? Again, look to the history of the civil rights struggle. After many decades of public denial and inaction, the civil rights movement helped Americans to see Southern apartheid in moral terms. From there, the movement succeeded by working toward legal change. Segregation was phased out rapidly only because it was phased out through the law. These statutes didn't erase racial prejudice from every American heart overnight. But through them, our country made staggering progress. Just consider who occupies the White House today.

All who appreciate the enormity of the climate crisis still have a responsibility to make every change possible in their personal lives. I have, from the solar panels on my roof to the Prius in my driveway to my low-carbon-footprint vegetarian diet. But surveys show that very few people are willing to make significant voluntary changes, and those of us who do create the false impression of mass progress as the media hypes our actions.

Instead, most people want carbon reductions to be mandated by laws that will allow us to share both the responsibilities and the benefits of change. Ours is a nation of laws; if we want to alter our practices in a deep and lasting way, this is where we must start. After years of delay and denial and green half-measures, we must legislate a stop to the burning of coal, oil and natural gas.

Of course, all this will require congressional action, and therein lies the source of Obama's Copenhagen headache. To have been in the strongest position to negotiate a binding emissions treaty with other world leaders this month, the president needed a strong carbon-cap bill out of Congress. But the House of Representatives passed only a weak bill riddled with loopholes in June, and the Senate has failed to get even that far.

So what's the problem? There's lots of blame to go around, but the distraction of the "go green" movement has played a significant role. Taking their cues from the popular media and cautious politicians, many Americans have come to believe that they are personally to blame for global warming and that they must fix it, one by one, at home. And so they either do as they're told—a little of this, a little of that—or they feel overwhelmed and do nothing.

We all got into this mess together. And now, with treaty talks underway internationally and Congress stalled at home, we need to act accordingly. Don't spend an hour changing your light bulbs. Don't take a day to caulk your windows. Instead, pick up a phone, open a laptop, or travel to a U.S. Senate office near you and turn the tables: "What are the ten green statutes you're working on to save the planet, Senator?"

Demand a carbon-cap bill that mandates the number 350. That's the level of carbon pollution scientists say we must limit ourselves to: 350 parts per million of $CO_2$ in the air. If we can stabilize the atmosphere at that number in coming decades, we should be able to avoid the worst-case scenario and preserve a planet similar to the one human civilization developed on. To get there, America will need to make deep but achievable pollution cuts well before 2020. And to protect against energy price shocks during this transition, Congress must include a system of direct rebates to consumers, paid for by auctioning permit fees to the dirty-energy companies that continue to pollute our sky.

Obama, too, needs to step up his efforts; it's not just Congress and the voters who have been misguided. Those close to the president say he understands the seriousness of global warming. But despite the issue's moral gravity, he's been paralyzed by political caution. He leads from the rear on climate change, not from the front.

Forty-five years ago, President Lyndon B. Johnson faced tremendous opposition on civil rights from a Congress dominated by Southern leaders, yet he spent the political capital necessary to answer a great moral calling. Whenever key bills on housing, voting, and employment stalled, he gave individual members of Congress the famous "Johnson treatment." He charmed. He pleaded. He threatened. He led, in other words. In person, and from the front.

Does anyone doubt that our charismatic current president has the capacity to turn up the heat? Imagine the back-room power of a full-on "Obama treatment" to defend America's flooding coastlines and burning Western forests. Imagine a two-pronged attack on the fickle, slow-moving Senate: Obama on one side and a tide of tweets and letters from voters like you.

So join me: put off the attic insulation job till January. Stop searching online for recycled gift-wrapping paper and sustainably farmed Christmas trees. Go beyond green fads for a month, and instead help make green history.

*Naomi Klein is best known as the author of* No Logo, *a book that powerfully attacks corporate globalization and rampant consumerism. The Canadian-born Klein has emerged as an important leader as well as an author, and in the run-up to the Copenhagen conference in December 2009 she sounded a series of powerful alarms, warning that the strategies for fighting global warming were turning into a series of gifts for big corporate interests. At the conference itself, she became an important voice for the NGOs shut out of the negotiating sessions, and made a convincing case that there was little chance of solving global warming in isolation from the other problems of power and poverty plaguing the globe.*

# Climate Rage

Naomi Klein
2009

One last chance to save the world—for months, that's how the United Nations summit on climate change in Copenhagen, which starts in early December, was being hyped. Officials from 192 countries were finally going to make a deal to keep global temperatures below catastrophic levels. The summit called for "that old comic-book sensibility of uniting in the face of a common danger threatening the Earth," said Todd Stern, President Obama's chief envoy on climate issues. "It's not a meteor or a space invader, but the damage to our planet, to our community, to our children and their children will be just as great."

That was back in March. Since then, the endless battle over health care reform has robbed much of the president's momentum on climate change. With Copenhagen now likely to begin before Congress has passed even a weak-ass climate bill co-authored by the coal lobby, U.S. politicians have dropped the superhero metaphors and are scrambling to lower expectations for achieving a serious deal at the climate summit. It's just one meeting, says U.S. Energy Secretary Steven Chu, not "the be-all and end-all."

As faith in government action dwindles, however, climate activists are treating Copenhagen as an opportunity of a different kind. On track to be the largest environmental gathering in history, the summit represents a chance to seize the political terrain back from business-friendly half-measures, such as carbon offsets and emissions trading, and introduce some effective, common-sense proposals—ideas that have less to do with creating complex new markets for pollution and more to do with keeping coal and oil in the ground.

Among the smartest and most promising—not to mention controversial—proposals is "climate debt," the idea that rich countries should pay reparations to poor countries for the climate crisis. In the world of climate-change activism, this marks a dramatic shift in both tone and content. American environmentalism tends to treat global warming as a force that transcends difference: We all share this fragile blue planet, so we all need to work together to save it. But the coalition of Latin American and African governments making the case for climate debt actually stresses difference, zeroing in on the cruel contrast between those who caused the climate crisis (the developed world) and those who are suffering its worst effects (the developing world). Justin Lin, chief economist at the World Bank, puts the equation bluntly: "About seventy-five to eighty percent" of the damages caused by global warming "will be suffered by developing countries, although they only contribute about one-third of greenhouse gases."

Climate debt is about who will pick up the bill. The grassroots movement behind the proposal argues that all the costs associated with adapting to a more hostile ecology—everything from building stronger sea walls to switching to cleaner, more expensive technologies—are the responsibility of the countries that created the crisis. "What we need is not something we should be begging for but something that is owed to us, because we are dealing with a crisis not of our making," says Lidy Nacpil, one of the coordinators of Jubilee South, an international organization that has staged demonstrations to promote climate reparations. "Climate debt is not a matter of charity."

Sharon Looremeta, an advocate for Maasai tribespeople in Kenya who have lost at least five million cattle to drought in recent years, puts it in even sharper terms. "The Maasai community does not drive 4x4s or fly off on holidays in airplanes," she says. "We have not caused climate change, yet we are the ones suffering. This is an injustice and should be stopped right now."

The case for climate debt begins like most discussions of climate change: with the science. Before the Industrial Revolution, the density of carbon dioxide in the atmosphere—the key cause of global warming—was about 280 parts per million. Today, it has reached 387 parts per million—far above safe limits—and it's still rising. Developed countries, which represent less than twenty percent of the world's population, have emitted almost seventy-five percent of all greenhouse gas pollution that is now destabilizing the climate. (The U.S. alone, which comprises barely five percent of the global population, contributes twenty-five percent of all carbon emissions.) And while developing countries like China and India have also begun to spew large amounts of carbon dioxide, the reasoning goes, they are not equally responsible for the cost of the cleanup, because they have contributed only a small fraction of the 200 years of cumulative pollution that has caused the crisis.

In Latin America, left-wing economists have long argued that Western powers owe a vaguely defined "ecological debt" to the continent for centuries of colonial land-grabs and resource extraction. But the emerging argument for climate debt is far more concrete, thanks to a relatively new body of research putting precise figures on who emitted what and when. "What is exciting," says Antonio Hill, senior climate adviser at Oxfam, "is you can really put numbers on it. We can measure it in tons of $CO_2$ and come up with a cost."

Equally important, the idea is supported by the United Nations Framework Convention on Climate Change—ratified by 192 countries, including the United States. The framework not only asserts that "the largest share of historical and current global emissions of greenhouse gases has originated in developed countries," it clearly states that actions taken to fix the problem should be made "on the basis of equity and in accordance with their common but differentiated responsibilities."

The reparations movement has brought together a diverse coalition of big international organizations, from Friends of the Earth to the World Council of Churches, that have joined up with climate scientists and political economists, many of them linked to the influential Third World Network, which has been leading the call. Until recently, however, there was no government pushing for climate debt to be included in the Copenhagen agreement. That changed in June, when Angelica Navarro, the chief climate negotiator for Bolivia, took the podium at a U.N. climate negotiation in Bonn, Germany. Only thirty-six and dressed casually in a black sweater, Navarro looked more like the hippies outside than the bureaucrats and civil servants inside the session. Mixing the latest emissions science with accounts of how melting glaciers were threatening the water supply in two major Bolivian cities, Navarro made the case for why developing countries are owed massive compensation for the climate crisis.

"Millions of people—in small islands, least-developed countries, landlocked countries as well as vulnerable communities in Brazil, India and China, and all around the world—are suffering from the effects of a problem to which they did not contribute," Navarro told the packed room. In addition to facing an increasingly hostile climate, she added, countries like Bolivia cannot fuel economic growth with cheap and dirty energy, as the rich countries did, since that would only add to the climate crisis—yet they cannot afford the heavy upfront costs of switching to renewable energies like wind and solar.

The solution, Navarro argued, is three-fold. Rich countries need to pay the costs associated with adapting to a changing climate, make deep cuts to their own emission levels "to make atmospheric space available" for the developing world, and pay Third World countries to leapfrog over fossil fuels and go straight to cleaner alternatives. "We cannot and will not give up our rightful claim to a fair share of atmospheric space on the promise that, at some future stage, technology will be provided to us," she said.

The speech galvanized activists across the world. In recent months, the governments of Sri Lanka, Venezuela, Paraguay and Malaysia have endorsed the concept of climate debt. More than 240 environmental and development organizations have signed a statement calling for wealthy nations to pay their climate debt, and forty-nine of the world's least-developed countries will take the demand to Copenhagen as a negotiating bloc.

"If we are to curb emissions in the next decade, we need a massive mobilization larger than any in history," Navarro declared at the end of her talk. "We need a Marshall Plan for the Earth. This plan must mobilize financing and technology transfer on scales never seen before. It must get technology onto the ground in every country to ensure we reduce emissions while raising people's quality of life. We have only a decade."

A very expensive decade. The World Bank puts the cost that developing countries face from climate change—everything from crops destroyed by drought and floods to malaria spread by mosquito-infested waters—as high as $100 billion a year. And shifting to renewable energy, according to a team of United Nations researchers, will raise the cost far more: to as much as $600 billion a year over the next decade.

Unlike the recent bank bailouts, however, which simply transferred public wealth to the world's richest financial institutions, the money spent on climate debt would fuel a global environmental transformation essential to saving the entire planet. The most exciting example of what could be accomplished is the ongoing effort to protect Ecuador's Yasuní National Park. This extraordinary swath of Amazonian rainforest, which is home to several indigenous tribes and a surreal number of rare and exotic animals, contains nearly as many species of trees in 2.5 acres as exist in all of North America. The catch is that underneath that riot of life sits an estimated 850 million barrels of crude oil, worth about $7 billion. Burning that oil—and logging the rainforest to get it—would add another 547 million tons of carbon dioxide to the atmosphere.

Two years ago, Ecuador's center-left president, Rafael Correa, said something very rare for the leader of an oil-exporting nation: He wanted to leave the oil in the ground. But, he argued, wealthy countries should pay Ecuador—where half the population lives in poverty—not to release that carbon into the atmosphere, as "compensation for the damages caused by the out-of-proportion amount of historical and current emissions of greenhouse gases." He didn't ask for the entire amount; just half. And he committed to spending much of the money to move Ecuador to alternative energy sources like solar and geothermal.

Largely because of the beauty of the Yasuní, the plan has generated widespread international support. Germany has already offered $70 million a year for thirteen years, and several other European governments have expressed interest in participating. If Yasuní is saved, it will demonstrate that climate debt isn't just a disguised ploy for more aid—it's a far more credible solution to the climate crisis than the ones we have now. "This initiative needs to succeed," says Atossa Soltani, executive director of Amazon Watch. "I think we can set a model for other countries."

Activists point to a huge range of other green initiatives that would become possible if wealthy countries paid their climate debts. In India, mini power plants that run on biomass and solar power could bring low-carbon electricity to many of the 400 million Indians currently living without a light bulb. In cities from Cairo to Manila, financial support could be given to the armies of impoverished "trash pickers" who save as much as eighty percent of municipal waste in some areas from winding up in garbage dumps and trash incinerators that release planet-warming pollution. And on a much larger scale, coal-fired power plants across the developing world could be converted into more efficient facilities using existing technology, cutting their emissions by more than a third.

But to ensure that climate reparations are real, advocates insist, they must be independent of the current system of international aid. Climate money cannot simply be diverted from existing aid programs, such as primary education or HIV prevention. What's more, the funds must be provided as grants, not loans, since the last thing developing countries need is more debt. Furthermore, the money should not be administered by the usual suspects like the World Bank and USAID, which too often push pet projects based on Western agendas, but must be controlled by the United Nations climate convention, where developing countries would have a direct say in how the money is spent.

Without such guarantees, reparations will be meaningless—and without reparations, the climate talks in Copenhagen will likely collapse. As it stands, the U.S. and other Western nations are engaged in a lose-lose game of chicken with developing nations like India and China: We refuse to lower our emissions unless they cut theirs and submit to international monitoring, and they refuse to budge unless wealthy nations cut first and cough up serious funding to help them adapt to climate change and switch to clean energy. "No money, no deal," is how one of South Africa's top environmental officials put it. "If need be," says Ethiopian Prime Minister Meles Zenawi, speaking on behalf of the African Union, "we are prepared to walk out."

In the past, President Obama has recognized the principle on which climate debt rests. "Yes, the developed nations that caused much of the damage to our climate over the last century still have a responsibility to lead," he acknowledged in his September speech at the United Nations. "We have a responsibility to provide the financial and technical assistance needed to help these [developing] nations adapt to the impacts of climate change and pursue low-carbon development."

Yet as Copenhagen draws near, the U.S. negotiating position appears to be to pretend that 200 years of over-emissions never happened. Todd Stern, the chief U.S. climate negotiator, has scoffed at a Chinese and African proposal that developed countries pay as much as $400 billion a year in climate financing as "wildly unrealistic" and "untethered to reality." Yet he put no alternative number on the table—unlike the European Union, which has offered to kick in up to $22 billion. U.S. negotiators have even suggested that countries could fund climate debt by holding periodic "pledge parties," making it clear that they see covering the costs of climate change as a matter of whimsy, not duty.

But shunning the high price of climate change carries a cost of its own. U.S. military and intelligence agencies now consider global warming a leading threat to national security. As sea levels rise and droughts spread, competition for food and water will only increase in many of the world's poorest nations. These regions will become "breeding grounds for instability, for insurgencies, for warlords," according to a 2007 study for the Center for Naval Analyses led by General Anthony Zinni, the former Centcom commander. To keep out millions of climate refugees fleeing hunger and conflict, a report commissioned by the Pentagon in 2003 predicted that the U.S. and other rich nations would likely decide to "build defensive fortresses around their countries."

Setting aside the morality of building high-tech fortresses to protect ourselves from a crisis we inflicted on the world, those enclaves and resource wars won't come cheap. And unless we pay our climate debt, and quickly, we may well find ourselves living in a world of climate rage. "Privately, we already hear the simmering resentment of diplomats whose countries bear the costs of our emissions," Senator John Kerry observed recently. "I can tell you from my own experience: It is real, and it is prevalent. It's not hard to see how this could crystallize into a virulent, dangerous, public anti-Americanism. That's a threat too. Remember: The very places least responsible for climate change—and least equipped to deal with its impacts—will be among the very worst affected."

That, in a nutshell, is the argument for climate debt. The developing world has always had plenty of reasons to be pissed off with their northern neighbors, with our tendency to overthrow their governments, invade their countries and pillage their natural resources. But never before has there been an issue so politically inflammatory as the refusal of people living in the rich world to make even small sacrifices to avert a potential climate catastrophe. In Bangladesh, the Maldives, Bolivia, the Arctic, our climate pollution is directly responsible for destroying entire ways of life—yet we keep doing it.

From outside our borders, the climate crisis doesn't look anything like the meteors or space invaders that Todd Stern imagined hurtling toward Earth. It looks, instead, like a long and silent war waged by the rich against the poor. And for that, regardless of what happens in Copenhagen, the poor will continue to demand their rightful reparations. "This is about the rich world taking responsibility for the damage done," says Ilana Solomon, policy analyst for ActionAid USA, one of the groups recently converted to the cause. "This money belongs to poor communities affected by climate change. It is their compensation."

*In the summer of 2010, the Senate refused even to vote on a mild bill that would have levied a small tax on electric utilities for their carbon emissions. The collapse of efforts by the mainline green groups in Washington, led by the corporate-friendly Environmental Defense Fund, showed that the energy industry was unwilling to make even small compromises to its business model. And so began a new chapter in the global warming fight. With tepid legislation off the table, and every sign that Congress would continue ignoring the problem in the years ahead, the field was cleared for new attempts at large-scale organizing, including a shift towards more militant civil disobedience. Whether those attempts will come in time to alter the physics and chemistry of the atmosphere remain to be seen. But come they will.*

# This Is Fucked Up—It's Time to Get Mad, and Then Busy

Bill McKibben
*2010*

Try to fit these facts together:

- According[1] to the National Oceanic and Atmospheric Administration, the planet has just come through the warmest decade, the warmest twelve months, the warmest six months, and the warmest April, May, and June on record.
- A "staggering" new study[2] from Canadian researchers has shown that warmer seawater has reduced phytoplankton, the base of the marine food chain, by forty percent since 1950.
- Nine nations[3] have so far set their all-time temperature records in 2010, including Russia (111 degrees), Niger (118), Sudan (121), Saudi Arabia and Iraq (126 apiece), and Pakistan, which also set the new all-time Asia record[4] in May: a hair under 130 degrees. I can turn my oven to 130 degrees.
- And then, in late July, the U.S. Senate decided to do exactly nothing about climate change. They didn't do less than they could have—they did nothing, preserving a perfect two-decade bipartisan record of no action. Senate majority leader Harry Reid decided not even to schedule a vote on legislation that would have capped carbon emissions.

I wrote the first book for a general audience on global warming back in 1989, and I've spent the subsequent twenty-one years working on the issue. I'm a mild-mannered guy, a Methodist Sunday School teacher. Not quick to anger. So what I want to say is: this is fucked up. The time has come to get mad, and then to get busy.

For many years, the lobbying fight for climate legislation on Capitol Hill has been led by a collection of the most corporate and moderate environmental groups, outfits like the Environmental Defense Fund. We owe them a great debt, and not just for their hard work. We owe them a debt because they did everything the way you're supposed to: they wore nice clothes, lobbied tirelessly, and compromised at every turn.

By the time they were done, they had a bill that only capped carbon emissions from electric utilities (not factories or cars) and was so laden with gifts for industry that if you listened closely you could actually hear the oinking. They bent over backwards like Soviet gymnasts. Senator John Kerry, the legislator they worked most closely with, issued this rallying cry as the final negotiations began: "We believe we have compromised significantly, and we're prepared to compromise further."

And even that was not enough. They were left out to dry by everyone—not just Reid, not just the Republicans. Even President Obama wouldn't lend a hand, investing not a penny of his political capital in the fight.

The result: total defeat, no moral victories.

## Now What?

So now we know what we didn't before: making nice doesn't work. It was worth a try, and I'm completely serious when I say

I'm grateful they made the effort, but it didn't even come close to working. So we better try something else.

Step one involves actually talking about global warming. For years now, the accepted wisdom in the best green circles was: talk about anything else—energy independence, oil security, beating the Chinese to renewable technology. I was at a session convened by the White House early in the Obama administration where some polling guru solemnly explained that "green jobs" polled better than "cutting carbon."

No, really? In the end, though, all these focus-group favorites are secondary. The task at hand is keeping the planet from melting. We need everyone—beginning with the president—to start explaining that basic fact at every turn.

It is the heat, and also the humidity. Since warm air holds more water than cold, the atmosphere is about five percent moister than it was forty years ago, which explains the freak downpours that seem to happen someplace on this continent every few days. It is the carbon—that's why the seas are turning acid, a point Obama could have made with ease while standing on the shores of the Gulf of Mexico. "It's bad that it's black out there," he might have said, "but even if that oil had made it safely ashore and been burned in our cars, it would still be wrecking the oceans." Energy independence is nice, but you need a planet to be energy independent on.

Mysteriously enough, this seems to be a particularly hard point for smart people to grasp. Even in the wake of the disastrous Senate non-vote, the Nature Conservancy's climate expert told[5] New York Times columnist Tom Friedman, "We have to take climate change out of the atmosphere, bring it down to earth, and show how it matters in people's everyday lives." Translation: ordinary average people can't possibly recognize the real stakes here, so let's put it in language they can understand, which is about their most immediate interests. It's both untrue, as I'll show below, and incredibly patronizing. It is, however, exactly what we've been doing for a decade and, clearly, It Does Not Work.

Step two, we have to ask for what we actually need, not what we calculate we might possibly be able to get. If we're going to slow global warming in the very short time available to us, then we don't actually need an incredibly complicated legislative scheme that gives door prizes to every interested industry and turns the whole operation over to Goldman Sachs[6] to run. We need a stiff price on carbon, set by the scientific understanding that we can't still be burning black rocks a couple of decades hence. That undoubtedly means upending the future business plans of Exxon and BP, Peabody Coal and Duke Energy, not to speak of everyone else who's made a fortune by treating the atmosphere as an open sewer for the byproducts of their main business.

Instead they should pay through the nose for that sewer, and here's the crucial thing: *most of the money raised in the process should be returned directly to American pockets.* The monthly check sent to Americans would help fortify us against the rise in energy costs, and we'd still be getting the price signal at the pump to stop driving that SUV and start insulating the house. We also need to make real federal investments in energy research and development, to help drive down the price of alternatives—the Breakthrough Institute points out,[7] quite rightly, that we're crazy to spend more of our tax dollars on research into new drone aircraft and Mars orbiters than we do on photovoltaics.

Yes, these things are politically hard, but they're not impossible. A politician who really cared could certainly use, say, the platform offered by the White House to sell a plan that taxed BP and actually gave the money to ordinary Americans. (So far they haven't even used the platform offered by the White House to reinstall[8] the rooftop solar panels that Jimmy Carter put there in the 1970s and Ronald Reagan took down in his term.)

Asking for what you need doesn't mean you'll get all of it. Compromise still happens. But as David Brower, the greatest environmentalist of the late twentieth century, explained[9] amid the fight to save the Grand Canyon: "We are to hold fast to what we believe is right, fight for it, and find allies and adduce all possible arguments for our cause. If we cannot find enough vigor in us or them to win, then let someone else propose the compromise. We thereupon work hard to coax it our way. We become a nucleus around which the strongest force can build and function."

Which leads to the third step in this process. If we're going to get any of this done, we're going to need a movement, the one thing we haven't had. For twenty years environmentalists have operated on the notion that we'd get action if we simply had scientists explain to politicians and CEOs that our current ways were ending the Holocene,[10] the current geological epoch. That turns out, quite conclusively, not to work. We need to be able to explain that their current ways will end something they actually care about, i.e. their careers. And since we'll never have the cash to compete with Exxon, we better work in the currencies we can muster: bodies, spirit, passion.

*Movement Time*

As Tom Friedman put it in a strong column[11] the day after the Senate punt, the problem was that the public "never got mobilized." Is it possible to get people out in the streets demanding action about climate change? Last year, with almost no money, our scruffy little outfit, 350.org,[12] managed to organize what *Foreign Policy* called[13] the "largest ever coordinated global rally of any kind" on any issue—5,200 demonstrations in 181 countries, 2,000 of them in the United States.

People were rallying not just about climate change, but around a remarkably wonky scientific data point, 350 parts per million carbon dioxide, which NASA's James Hansen and his colleagues have demonstrated[14] is the most we can have in the atmosphere if we want a planet "similar to the one on which civilization developed and to which life on earth is adapted." Which, come to think of it, we do. And the "we," in this case, was not rich white folks. If you look at the 25,000 pictures in our Flickr account,[15] you'll see that most of them were poor, black, brown, Asian, and young—because that's what most of the world is. No need for vice presidents of big conservation groups to patronize them: shrimpers in Louisiana and women in burqas and priests in Orthodox churches and slumdwellers in Mombasa turned out to be completely capable of understanding the threat to the future.

Those demonstrations were just a start (one we should have made long ago). We're following up in October on 10-10-10— with a Global Work Party.[16] All around the country and the world people will be putting up solar panels and digging community gardens and laying out bike paths. Not because we can stop climate change one bike path at a time, but because we need to make a sharp political point to our leaders: we're getting to work, what about you?

We need to shame them, starting now. And we need everyone working together. This movement is starting to emerge on many fronts. In September, for instance, opponents of mountaintop removal are converging on D.C.[17] to demand an end to the coal trade. That same month, Tim DeChristopher goes on trial in Salt Lake City for monkey-wrenching oil and gas auctions by submitting phony bids. (Naomi Klein and Terry Tempest Williams have called for folks to gather at[18] the courthouse.)

The big environmental groups are starting to wake up, too. The Sierra Club has a dynamic new leader, Mike Brune,[19] who's working hard with stalwarts like Greenpeace and Friends of the Earth. (Note to enviro groups: working together is fun and useful.) Churches[20] are getting involved, as well as mosques and synagogues. Kids are leading[21] the fight all over the world—they have to live on this planet for another seventy years or so, and they have every right to be pissed off.

*But no one will come out to fight for watered-down and weak legislation.* That's not how it works. You don't get a movement unless you take the other two steps I've described.

And in any event it won't work overnight. We're not going to get the Senate to act next week, or maybe even next year. It took a decade after the Montgomery bus boycott to get the Voting Rights Act. But if there hadn't been a movement, then the Voting Rights Act would have passed in… never. We may need to get arrested. We definitely need art, and music, and disciplined, nonviolent, but very real anger.

Mostly, we need to tell the truth, resolutely and constantly. Fossil fuel is wrecking the one earth we've got. It's not going to go away because we ask politely.

If we want a world that works, we're going to have to raise our voices.

*Adrienne Maree Brown here captures some of the most important voices in the fight against climate change: those of young people, who will live with the effects of inaction far longer than most of the aging political leaders making the decisions. (Popular T-shirt among youth in Copenhagen: "How old will you be in 2050?") In one poll after another, young people demonstrate a firmer grasp of the science of global warming, and a much higher interest in taking action, than their elders: among young supporters of President Obama in 2008, for instance, global warming was the most salient issue, yet it hardly made the list for senior citizens. The climate change movement around the world is powered largely by young people, which partly explains why it has been such an early user of Facebook, Twitter, and similar technologies. Adrienne Maree Brown, by the way, is national coordinator of the U.S. Social Forum, and a leader of the wonderfully (and aptly) named Ruckus Society.*

# The Green Generation

Adrienne Maree Brown
*2007*

## Social Climate Change

There has been one constant, as far back as we can understand:
The world is always changing. Changing its surface, weather,
people and values. It happens over massive periods of time, and
it happens in the course of a generation, simultaneously, everywhere. It's hard not to assign blame for those shifts, especially
if some changes aren't survivable for humans.

Most of the world, including a privileged portion of the
United States, is now aware of global warming. Many populations in our country are already experiencing its effects. Today's
generation of sixteen- to thirty-five-year-olds experiences the
climate crisis in a number of ways.

This is the generation that will uncover the answers to
our biggest questions: how will we all survive the peak of our
resources? Will everyone make it? How does a planet move
from consumption to sustainability? Will we make the connection between global warming and economic disparity on a large
enough scale? So far, the only thing incremental about climate
change is our American response to it.

## Green Beginnings

By necessity, this is the generation that will produce business-people who recognize the longevity of their success as tied to green and socially just entrepreneurial endeavors. This generation will produce funders who strategically resource local work with an eye toward sustaining a whole system of strong communities, academics who research policy that values sustainability and justice, scientists and engineers who develop new green technologies that employ community members. This generation is teeming with activists and organizers who will make the big visionary connection and take the drastic strategic actions on behalf of our survival.

Some youth experience it directly—in the Arctic and the Gulf Coast, youth have seen drastic changes in their land, homes, infrastructure, weather. In urban industrial and post-industrial areas like Detroit, New York, Chicago and Pittsburgh, youth experience the physical buildup of pollution in their bodies. Some experience it less directly—targeted heavily by military recruiters desperate to beef up forces for wars to secure our prolonged reliance on outdated resources.

I wanted to know what young people think about global warming. We are in a unique position as the first generation to have such complete access to information and to communications from all over the world, with such specific knowledge about huge changes happening around us. Who is aware of that bigger picture? Who is already doing work around it?

I spoke to a wide variety of young people—youth in every city I traveled to this summer, youth who identify as Christian, Muslim and atheist. Youth from public schools and private schools, youth who were home-schooled, youth I surveyed on

MySpace and Facebook, youth from the Gulf Coast and the arctic, youth who consider themselves environmentalists and youth who don't.

*Wiretap:* Here, in their own words, the Green Generation.

I could go on forever about what I would do to stop global warming. —*Leah*

*Wiretap:* What is global warming?

Earth temperatures consistently rising faster than any natural cycle due to human pollution. —*Wini (16, Brooklyn, New York)*

The earth is getting hotter (e.g., average temperatures are rising around the world, and there is a subsequent shift in worldwide weather patterns) because of all the shit we—humans—put into the atmosphere. —*Leah (25, Texas)*

Can't explain it… the rain forest cutting down has a lot to do with it, changing the weather and taking away natural medications. —*Edy (16, Brooklyn, New York)*

Uh… human activity fucking (can I say that?) with the atmosphere and environment and causing a very small but detrimental increase in the earth's temperature. — *Erin (23, Northwest Territories, Canada)*

A phenomenon where greenhouse gases emitted from car exhaust, coal furnaces and other pollutants build up in the atmosphere and create more cloud cover and a denser atmosphere where light and heat from the sun bounces back and forth between the earth and the ozone instead of escaping. The buildup of light and heat cause the temperatures within the atmosphere to increase over time, which can cause polar-ice-cap melting, glacial melts, El Niño/La Niña phenomenon, erratic weather patterns, etc. —*Lisa (16, Selma, Alabama)*

Something when the ice melts and overflows the world, and it could kill us! —*Darnell (16)*

Global warming… a relatively abrupt increase in the average temperature of the globe. The current widely accepted theory is that this is the result of an increase in greenhouse gases in the atmosphere from human and natural sources. Greenhouse gases, such as carbon dioxide, methane, and CFCs (which can have one hundred times the effect of $CO_2$) absorb light from the sun and convert this energy into vibrations and collisions observed as heat. In an excess of normal levels of these greenhouse gases, more light energy is converted to heat energy, leading to an increase in the average global temperature. —*Dave*

To me, global warming is mankind destroying Mother Earth. Everything that was created by mankind to make life easier is only making it harder on Mother Earth. In turn, we are getting what is called global warming, which is the earth's refrigerator shutting down and the world pretty much collapsing. —*Kyla (21, Northwest Territories, Canada)*

*Wiretap:* How did you learn about global warming?

I have learned about this in high school, college and graduate school as well as through lectures and the media. —*Dave*

Elementary school in the '80s. Was I the only one? —*Wini (16, Brooklyn, New York )*

On *Oprah* Al Gore was talking about it yesterday. —*Pam (16, Selma, Alabama)*

Didn't learn about it! —*Arthur (19, Selma, Alabama)*

*Wiretap:* Do you think we will see the effects of global warming in your lifetime?

We have been seeing the effects of global warming already. Take a look at old and new pictures of ancient glaciers to see the stark difference. Look at weather phenomena like Hurricane Katrina and Rita. The bees are disappearing, which some people think could be the canary in the coal mine. Polar bears are dying because their icebergs melt beneath their feet. Wine country in Napa Valley, California, and Bordeaux, France, is drying up and more northern wine crops are yielding better. Global warming has been happening since the Industrial Revolution. — *Lisa (16, Selma, Alabama)*

I think we're seeing them already! —*Casey (16, North Augusta, South Carolina)*

Yes. I want to write the book *100 Places to Go Before They Die.* Just go to Yosemite, turn on the Weather

Channel, visit the Canadian ice fields or walk outside to see the effects. —*Lauren (16, Selma, Alabama)*

I believe we are going to experience catastrophic effects of global warming in my lifetime. These may include mass species extinction, habitat destruction, rising sea levels, and astonishing (and dangerous) shifts in global weather patterns. —*Leah (25, Texas)*

I am seeing it today. Living in the north, global warming is all around me. The banks are eroding, the water is rising, the winters are shorter and summers longer. The animals are adapting to this situation, which means that we are seeing animals in our areas that don't belong (salmon in the Mackenzie delta; seals, otters in the Mackenzie islands; polar and grizzly bears, and so on).
—*Kyla (21, Northwest Territories, Canada)*

*Wiretap:* Is your generation doing enough?

I think people my age have little hope that we can do anything as individuals to really effect change. There are people who are wildly inspired and do everything they can to live a more sustainable lifestyle. Then there are people in the middle, like me. And then there are people who are clueless. I think most people are clueless.
—*Wini (16, Brooklyn, New York)*

There is not much you do on a personal level.
—*Edy (16, Brooklyn, New York)*

No, not yet. We are reactionary and will try to make bigger changes probably way too late. —*Molasses*

Most people do not do anything to stop global warming. People don't even recycle when the bins are right next to the trash. People do not choose cars for their fuel efficiency. People do not work so they can change the world—most people I've met work so they can make money and buy a new car or a new TV. It is not a popular position to be an idealist, and most people, regardless of their age, think only about their own happiness today.
—*Lisa (16, Selma, Alabama)*

People my age are not doing enough, because enough would mean putting enough pressure on the older status quo generation running the mechanisms of power, so that we could see sound environmental policies trump corporatism. —*Gregory (32, Brooklyn, New York)*

Everyone feels like their own action isn't going to be enough to prevent it anyway, so why try—but of course, if everyone did change, it may just be enough. But people in their twenties seem to be very focused on themselves and their careers and maybe slightly less so on world issues.
—*Erin (23, Northwest Territories, Canada)*

Unless you work for a global warming prevention initiative, ride a bike everywhere, farm all your own food, and live in a teepee with a lamp that runs on wind power, you are probably not doing enough. I actually feel so strongly about the coming changes in our world that I have completely switched careers so that I am armed to meet these changes with hard skills rather than just the ability to talk and write about policy. (I do think those

skills are also, of course, extremely important, but when the shit hits the fan, I want to really be able to help!) That is just one of many reasons I have switched careers and am now working toward becoming a doctor.
—*Leah (25, Texas)*

No. However, I don't think that any age group is doing enough to stop global warming. —*Alex*

*Wiretap:* Even as they are aware that not enough is being done, most of the folks I interviewed are engaged in some actions of their own.

I already have one solar panel. And I plan on covering my south-facing roof with many more. I currently drive a car that gets thirty-seven miles to the gallon. When the technology improves, my next car will likely be electric unless something better comes along in five years. We recycle! We buy the green option from the power company so extra money goes towards renewable energies currently being used to generate electricity.
—*Lisa (16, Selma, Alabama)*

I do try to reduce my carbon footprint by biking instead of driving, eating more locally grown food, etc. —*Alex*

I am committed to greening the entertainment industry. This is my large-scale move to stop global warming, but on a small scale, I walk/bike whenever possible and lead people on hikes to inform them about the effects of global warming and mismanagement of the environment.
—*Lauren (16, Selma, Alabama)*

*The Future Is Now*

What occurred to me reviewing the answers was that any lack of knowledge, any naïveté, any failures on the part of the public education system, media, and parents to inform this generation of its environmental circumstances—it's all temporary. We might be learning slowly, but change is coming fast.

> I think it is possible to stop global warming, one person at a time. —*Lisa (16, Selma, Alabama)*

Or 8,000 people. Or 17,000 people. Or a couple million people. In the course of writing this, I have come across young people at gatherings like Bioneers, Power Shift 2007, and the Brower Youth Awards. In each space there is more action than I've ever seen before by young people from impacted communities who have the privilege of being awake to what's happening and are looking for people to move the front line of their sustainable visions forward against all odds.

*George Monbiot is a British journalist and activist whose work often appears in the Guardian. His writing is notable not only for its felicity but for its fearlessness, evident in this piece where he engages an oft-heard truism about climate change: that "population" is the underlying issue. It's true that the pressure exerted by a growing human population contributes to our troubles—but advances in educating and empowering women have actually brought birth rates down, and, as Monbiot points out, most of the increase still expected before our numbers level out at midcentury will occur among people who use so little energy they won't generate much carbon. A far greater problem is rising consumption in places like China, which have relatively stable, but increasingly affluent populations. Of course, they're only emulating the consumption and carbon champs of the West.*

# The Population Myth

George Monbiot
2009

It's no coincidence that most of those who are obsessed with population growth are post-reproductive wealthy white men: it's about the only environmental issue for which they can't be blamed. The brilliant earth-systems scientist James Lovelock, for example, claimed last month that "Those who fail to see that population growth and climate change are two sides of the same coin are either ignorant or hiding from the truth. These two huge environmental problems are inseparable and to discuss one while ignoring the other is irrational."[1] But it's Lovelock who is being ignorant and irrational.

A paper published yesterday in the journal *Environment and Urbanization* shows that the places where population has been growing fastest are those in which carbon dioxide has been growing most slowly, and vice versa. Between 1980 and 2005, for example, sub-Saharan Africa produced 18.5 percent of the world's population growth and just 2.4 percent of the growth in $CO_2$. North America turned out four percent of the extra people, but fourteen percent of the extra emissions. Sixty-three percent of the world's population growth happened in places with very low emissions.[2]

Even this does not capture it. The paper points out that around one sixth of the world's population is so poor that it produces no significant emissions at all. This is also the group whose growth rate is likely to be highest. Households in India earning less than 3,000 rupees a month use a fifth of the electricity per head and one seventh of the transport fuel of households earning Rs30,000 or more. Street sleepers use almost nothing. Those who live by processing waste (a large part of the urban underclass) often save more greenhouse gases than they produce.

Many of the emissions for which poorer countries are blamed should in fairness belong to us. Gas flaring by companies exporting oil from Nigeria, for example, has produced more greenhouse gases than all other sources in sub-Saharan Africa put together.[3] Even deforestation in poor countries is driven mostly by commercial operations delivering timber, meat, and animal feed to rich consumers. The rural poor do far less harm.[4]

The paper's author, David Satterthwaite of the International Institute for Environment and Development, points out that the old formula taught to all students of development—that total impact equals population times affluence times technology (I=PAT)—is wrong. Total impact should be measured as I=CAT: consumers times affluence times technology. Many of the world's people use so little that they wouldn't figure in this equation. They are the ones who have the most children.

While there's a weak correlation between global warming and population growth, there's a strong correlation between global warming and wealth. I've been taking a look at a few super-yachts, as I'll need somewhere to entertain Labour ministers in the style to which they're accustomed. First I went through the plans for Royal Falcon Fleet's RFF135, but when I discovered that it burns only 750 litres of fuel per hour[5] I realized that it wasn't going to impress Lord Mandelson. I might raise half an eyebrow in Brighton with the Overmarine Mangusta 105, which sucks up 850 litres per hour.[6] But the raft that's really caught my eye is made by Wally Yachts in Monaco. The WallyPower 118 (which gives total wallies a sensation of power) consumes 3400 litres per hour when travelling at sixty knots.[7] That's nearly one litre per second. Another way of putting it is 31 litres per kilometre.[8]

Of course to make a real splash I'll have to shell out on teak and mahogany fittings, carry a few jet skis and a mini-submarine, ferry my guests to the marina by private plane and helicopter, offer them bluefin tuna sushi and beluga caviar and drive the beast so fast that I mash up half the marine life of the Mediterranean. As the owner of one of these yachts I'll do more damage to the biosphere in ten minutes than most Africans inflict in a lifetime. Now we're burning, baby.

Someone I know who hangs out with the very rich tells me that in the banker belt of the lower Thames valley there are people who heat their outdoor swimming pools to bath temperature, all round the year. They like to lie in the pool on winter nights, looking up at the stars. The fuel costs them £3,000 a month. One hundred thousand people living like these bankers would knacker our life support systems faster than ten billion people living like the African peasantry. But at least the super-wealthy have the good manners not to breed very much, so the rich old men who bang on about human reproduction leave them alone.

In May the Sunday *Times* carried an article headlined "Billionaire club in bid to curb overpopulation." It revealed that "some of America's leading billionaires have met secretly" to decide which good cause they should support. "A consensus emerged that they would back a strategy in which population growth would be tackled as a potentially disastrous environmental, social and industrial threat."[9] The ultra-rich, in other words, have decided that it's the very poor who are trashing the planet. You grope for a metaphor, but it's impossible to satirise.

James Lovelock, like Sir David Attenborough and Jonathan Porritt, is a patron of the Optimum Population Trust (OPT). It is one of dozens of campaigns and charities whose sole purpose is to discourage people from breeding in the name of saving the biosphere. But I haven't been able to find any campaign whose sole purpose is to address the impacts of the very rich.

The obsessives could argue that the people breeding rapidly today might one day become richer. But as the super-wealthy grab an ever greater share and resources begin to run dry, this, for most of the very poor, is a diminishing prospect. There are strong social reasons for helping people to manage their reproduction, but weak environmental reasons, except among wealthier populations.

The Optimum Population Trust glosses over the fact that the world is going through demographic transition: population growth rates are slowing down almost everywhere and the number of people is likely, according to a paper in *Nature*, to peak this century,[10] probably at around ten billion.[11] Most of the growth will take place among those who consume almost nothing.

But no one anticipates a consumption transition. People breed less as they become richer, but they don't consume less; they consume more. As the habits of the super-rich show, there are no limits to human extravagance. Consumption can be expected to rise with economic growth until the biosphere hits the buffers. Anyone who understands this and still considers that population, not consumption, is the big issue is, in Lovelock's words, "hiding from the truth." It is the worst kind of paternalism, blaming the poor for the excesses of the rich.

So where are the movements protesting about the stinking rich destroying our living systems? Where is the direct action against super-yachts and private jets? Where's class war when you need it?

It's time we had the guts to name the problem. It's not sex; it's money. It's not the poor; it's the rich.

*James Hansen has been the most important scientist of the global warming era, and in recent years he's become one of its most outspoken political leaders as well. On the twentieth anniversary of his 1988 testimony before Congress, Hansen takes stock of where our country stands on climate change and of how much work we have left to do if we are to avert the disasters predicted for the future. Scorning the so-called cap-and-trade approach favored by the most business-oriented of American green groups, Hansen calls instead for a more straightforward tax that would put the coal industry out of business as quickly as possible. And citing the tendency of politics to get in the way of real change, he exhorts the American public to take matters into our own hands.*

# Global Warming Twenty Years Later: Tipping Points Near

James Hansen
*2008*

My presentation today is exactly twenty years after my 23 June 1988 testimony to Congress, which alerted the public that global warming was underway. There are striking similarities between then and now, but one big difference.

Again a wide gap has developed between what is understood about global warming by the relevant scientific community and what is known by policymakers and the public. Now, as then, frank assessment of scientific data yields conclusions that are shocking to the body politic. Now, as then, I can assert that these conclusions have a certainty exceeding ninety-nine percent.

The difference is that now we have used up all slack in the schedule for actions needed to defuse the global warming time bomb. The next president and Congress must define a course next year in which the United States exerts leadership commensurate with our responsibility for the present dangerous situation.

---

Dr. James E. Hansen, a physicist by training, directs the NASA Goddard Institute for Space Studies, a laboratory of the Goddard Space Flight Center and a unit of the Columbia University Earth Institute, but he speaks as a private citizen today at the National Press Club and at a Briefing to the House Select Committee on Energy Independence & Global Warming.

Otherwise it will become impractical to constrain atmospheric carbon dioxide, the greenhouse gas produced in burning fossil fuels, to a level that prevents the climate system from passing tipping points that lead to disastrous climate changes that spiral dynamically out of humanity's control.

Changes needed to preserve creation, the planet on which civilization developed, are clear. But the changes have been blocked by special interests, focused on short-term profits, who hold sway in Washington and other capitols.

I argue that a path yielding energy independence and a healthier environment is, barely, still possible. It requires a transformative change of direction in Washington in the next year.

On 23 June 1988 I testified to a hearing, chaired by Senator Tim Wirth of Colorado, that the Earth had entered a long-term warming trend and that human-made greenhouse gases almost surely were responsible. I noted that global warming enhanced both extremes of the water cycle, meaning stronger droughts and forest fires, on the one hand, but also heavier rains and floods.

My testimony two decades ago was greeted with skepticism. But while skepticism is the lifeblood of science, it can confuse the public. As scientists examine a topic from all perspectives, it may appear that nothing is known with confidence. But from such broad, open-minded study of all data, valid conclusions can be drawn.

My conclusions in 1988 were built on a wide range of inputs from basic physics, planetary studies, observations of ongoing changes, and climate models. The evidence was strong enough that I could say it was time to "stop waffling." I was sure that time would bring the scientific community to a similar consensus, as it has.

While international recognition of global warming was swift, actions have faltered. The U.S. refused to place limits on its emissions, and developing countries such as China and India rapidly increased their emissions.

What is at stake? Warming so far, about two degrees Fahrenheit over land areas, seems almost innocuous, being less than day-to-day weather fluctuations. But more warming is already "in the pipeline," delayed only by the great inertia of the world ocean. And climate is nearing dangerous tipping points. Elements of a "perfect storm," a global cataclysm, are assembled.

Climate can reach points such that amplifying feedbacks spur large rapid changes. Arctic sea ice is a current example. Global warming initiated sea-ice melt, exposing darker ocean that absorbs more sunlight, melting more ice. As a result, without any additional greenhouse gases, the Arctic soon will be ice-free in the summer.

More ominous tipping points loom. West Antarctic and Greenland ice sheets are vulnerable to even small additional warming. These two-mile-thick behemoths respond slowly at first, but if disintegration gets well underway it will become unstoppable. Debate among scientists is only about how much sea level would rise by a given date. In my opinion, if emissions follow a business-as-usual scenario, sea level rise of at least two meters is likely this century. Hundreds of millions of people would become refugees. No stable shoreline would be reestablished in any time frame that humanity can conceive.

Animal and plant species are already stressed by climate change. Polar and alpine species will be pushed off the planet, if warming continues. Other species attempt to migrate, but as some are extinguished their interdependencies can cause ecosystem collapse. Mass extinctions, of more than half the species on the planet, have occurred several times when the Earth warmed as much as expected if greenhouse gases continue to increase. Biodiversity recovered, but it required hundreds of thousands of years.

The disturbing conclusion, documented in a paper[1] I have written with several of the world's leading climate experts, is that the safe level of atmospheric carbon dioxide is no more than 350 ppm (parts per million) and it may be less. Carbon dioxide amount is already 385 ppm and rising about two ppm per year. Stunning corollary: the oft-stated goal to keep global warming less than two degrees Celsius (3.6 degrees Fahrenheit) is a recipe for global disaster, not salvation.

These conclusions are based on paleoclimate data showing how the Earth responded to past levels of greenhouse gases and on observations showing how the world is responding to today's carbon dioxide amount. The consequences of continued increase of greenhouse gases extend far beyond extermination of species and future sea level rise.

Arid subtropical climate zones are expanding poleward. Already an average expansion of about 250 miles has occurred, affecting the southern United States, the Mediterranean region, Australia and southern Africa. Forest fires and drying-up of lakes will increase further unless carbon dioxide growth is halted and reversed.

Mountain glaciers are the source of fresh water for hundreds of millions of people. These glaciers are receding world-wide, in the Himalayas, Andes and Rocky Mountains. They will disappear, leaving their rivers as trickles in late summer and fall, unless the growth of carbon dioxide is reversed.

Coral reefs, the rainforest of the ocean, are home for one-third of the species in the sea. Coral reefs are under stress for several reasons, including warming of the ocean, but especially because of ocean acidification, a direct effect of added carbon dioxide. Ocean life dependent on carbonate shells and skeletons is threatened by dissolution as the ocean becomes more acid.

Such phenomena, including the instability of Arctic sea ice and the great ice sheets at today's carbon dioxide amount, show that we have already gone too far. We must draw down atmospheric carbon dioxide to preserve the planet we know. A level of no more than 350 ppm is still feasible, with the help of reforestation and improved agricultural practices, but just barely—time is running out.

Requirements to halt carbon dioxide growth follow from the size of fossil carbon reservoirs.

Coal towers over oil and gas. Phase out of coal use, except where the carbon is captured and stored below ground, is the primary requirement for solving global warming.

Oil is used in vehicles where it is impractical to capture the carbon. But oil is running out. To preserve our planet we must also ensure that the next mobile energy source is not obtained by squeezing oil from coal, tar shale or other fossil fuels.

Fossil fuel reservoirs are finite, which is the main reason that prices are rising. We must move beyond fossil fuels eventually. Solution of the climate problem requires that we move to carbon-free energy promptly.

Special interests have blocked the transition to our renewable energy future. Instead of moving heavily into renewable energies, fossil companies choose to spread doubt about global warming, as tobacco companies discredited the smoking-cancer link. Methods are sophisticated, including funding to help shape school textbook discussions of global warming.

CEOs of fossil energy companies know what they are doing and are aware of long-term consequences of continued business as usual. In my opinion, these CEOs should be tried for high crimes against humanity and nature.

Conviction of ExxonMobil and Peabody Coal CEOs will be no consolation if we pass on a runaway climate to our children.

Humanity would be impoverished by ravages of continually shifting shorelines and intensification of regional climate extremes. Loss of countless species would leave a more desolate planet.

If politicians remain at loggerheads, citizens must lead. We must demand a moratorium on new coal-fired power plants. We must block fossil fuel interests who aim to squeeze every last drop of oil from public lands, off-shore, and wilderness areas. Those last drops are no solution. They yield continued exorbitant profits for a short-sighted self-serving industry, but no alleviation of our addiction or long-term energy source.

Moving from fossil fuels to clean energy is challenging, yet transformative in ways that will be welcomed. Cheap, subsidized fossil fuels engendered bad habits. We import food from halfway around the world, for example, even with healthier products available from nearby fields. Local produce would be competitive if not for fossil fuel subsidies and the fact that climate change damages and costs, due to fossil fuels, are also borne by the public.

A price on emissions that cause harm is essential. Yes, a carbon tax. Carbon tax with one hundred percent dividend[2] is needed to wean us off fossil fuel addiction. Tax and dividend allows the marketplace, not politicians, to make investment decisions.

Carbon tax on coal, oil and gas is simple, applied at the first point of sale or port of entry. The entire tax must be returned to the public, an equal amount to each adult, a half-share for children. This dividend can be deposited monthly in an individual's bank account.

Carbon tax with 100 percent dividend is non-regressive. On the contrary, you can bet that low and middle income people will find ways to limit their carbon tax and come out ahead. Profligate energy users will have to pay for their excesses.

Demand for low-carbon high-efficiency products will spur innovation, making our products more competitive on international markets. Carbon emissions will plummet as energy efficiency and renewable energies grow rapidly. Black soot, mercury and other fossil fuel emissions will decline. A brighter, cleaner future, with energy independence, is possible.

Washington likes to spend our tax money line-by-line. Swarms of high-priced lobbyists in alligator shoes help Congress decide where to spend, and in turn the lobbyists' clients provide "campaign" money.

The public must send a message to Washington. Preserve our planet, creation, for our children and grandchildren, but do not use that as an excuse for more tax-and-spend. Let this be our motto: "One hundred percent dividend or fight!"

The next president must make a national low-loss electric grid an imperative. It will allow dispersed renewable energies to supplant fossil fuels for power generation. Technology exists for direct-current high-voltage buried transmission lines. Trunk lines can be completed in less than a decade and expanded analogous to interstate highways.

Government must also change utility regulations so that profits do not depend on selling ever more energy, but instead increase with efficiency. Building code and vehicle efficiency requirements must be improved and put on a path toward carbon neutrality.

The fossil industry maintains its stranglehold on Washington via demagoguery, using China and other developing nations as scapegoats to rationalize inaction. In fact, we produced most of the excess carbon in the air today, and it is to our advantage as a nation to move smartly in developing ways to reduce emissions. As with the ozone problem, developing countries can be allowed limited extra time to reduce emissions. They will cooperate: they have much to lose from climate change and much to gain from clean air and reduced dependence on fossil fuels.

We must establish fair agreements with other countries. However, our own tax and dividend should start immediately. We have much to gain from it as a nation, and other countries will copy our success. If necessary, import duties on products from uncooperative countries can level the playing field, with the import tax added to the dividend pool.

Democracy works, but sometimes churns slowly. Time is short. The 2008 election is critical for the planet. If Americans turn out to pasture the most brontosaurian congressmen, if Washington adapts to address climate change, our children and grandchildren can still hold great expectations.

*Mohamed Nasheed, president of the Maldive Islands, delivered this speech early in the Copenhagen climate conference of December 2009. He was the first world leader to arrive, and—outspoken, fiery, and handsome—he was followed everywhere by camera crews. Nasheed is like a combination of Nelson Mandela and Barack Obama—several times a political prisoner of the Maldivian autocracy, he forced a free election in 2008, which he won handily, bringing a new spirit of openness to the country. But the world would have paid his country little attention if not for this one fact: the highest point in this archipelago of a thousand islands spread across the Indian Ocean is only a few meters above sea level. Very few countries face a more immediate threat from climate change, and none have acted as forcefully to protest inaction by the West (Nasheed taught his whole cabinet to scuba dive so they could hold an underwater cabinet meeting, demanding a 350 parts per million $CO_2$ target from the U.N.) and to show the way forward: though poor, the Maldive Islands have committed to becoming carbon neutral by 2020.*

# Speech at Klimaforum

Mohamed Nasheed
*2009*

Mr. McKibben, fellow environmentalists, ladies and gentlemen.
Four years ago myself and many fellow activists sat in solitary
confinement in Maldivian prison cells. We sat in those jail cells
not because we had committed any wrong. We sat in those cells
because we had deliberately broken the unjust laws of dictator-
ship. We had spoken out for a cause in which we believed. That
cause was freedom and democracy.

There were times, sitting in that prison, when I felt more
alone than you can imagine. There were times when I started to
believe the doubters, who said the Maldives would never become
free. Sometimes it felt like the doubters were right. The dictator-
ship had the guns, bombs and tanks. We had no weapons other
than the power of our words, and the moral clarity of our cause.
Many democracy activists like us had vanished, forgotten by
history, their struggle a failure.

But, in spite of the odds, we refused to give up hope. We
refused to listen to the voices of doubt and discouragement. We
refused to be swayed by those who could not see that change was
on the way. And we were right to stand up for what we believed.

We won our battle for democracy in the Maldives. I stand before you today as the first democratically elected president in the history of my country.

The path to democracy in the Maldives was not straightforward. It was bumpy and full of turns. But we were determined that no matter how difficult the terrain, we would reach the end of the road. And we succeeded in our cause.

Four years later and a continent away, we meet here to confront another seemingly impossible task. We are here to save our planet from the silent, patient and invisible enemy that is climate change.

And just as there were doubters in the Maldives, so there are doubters in Copenhagen. There are those who tell us that solving climate change is impossible. There are those who tell us taking radical action is too difficult. There are those who tell us to give up hope.

Well, I am here to tell you that we refuse to give up hope. We refuse to be quiet. We refuse to believe that a better world isn't possible.

I have three words to say to the doubters and deniers. Three words with which to win this battle. Just three words are all I need. You may already have heard them. *Three... Five... O. Three... Five... O.*

Three-Five-O saves the coral reefs. Three-Five-O keeps the Arctic frozen. Three-Five-O ensures my country survives. Three-Five-O makes a better world possible.

I am here to tell you that down the road in the Bella Center the Maldives team is fighting to keep Three-Five-O in the negotiating text.

They need all the help they can get from you. Please keep supporting them.

And the good news is that we are now part of a growing bloc of nations, all committed to keeping Three-Five-O as the central guiding goal of our global survival plan.

These nations need your help and support too.

I am not a scientist, but I know that one of the laws of physics is that you cannot negotiate with the laws of physics. Three-Five-O is a law of atmospheric physics. You cannot cut a deal with Mother Nature. And we don't intend to try.

This is why, in March, the Maldives announced plans to become the first carbon-neutral country in the world. We intend to become carbon neutral in ten years. We will switch from oil to 100 percent renewable energy. And we will offset aviation pollution until a way can be found to decarbonize air transport too.

For us, going carbon neutral is not just the right thing to do. We believe it is also in our economic self-interest. Countries that have the foresight to green their economies today will be the winners of tomorrow. These pioneering countries will free themselves from the unpredictable price of foreign oil. They will capitalize on the new, green economy of the future. And they will enhance their moral standing, giving them greater political influence on the world stage. In the Maldives, we have relinquished our claim to high-carbon growth.

After all, it is not carbon we want, but development. It is not coal we want, but electricity. It is not oil we want, but transport. Low-carbon technologies now exist to deliver all the goods and services we need. Let us make the goal of using them.

Let us make the goal of reaching that all-important number: Three-Five-O.

We believe that if the Maldives can become carbon neutral; richer, larger countries can follow. But if there is one thing I know about politicians, it's that they won't act until their electorates act first. This is where you come in.

History shows us the power of peaceful protest. From the civil rights movement, to Gandhi's Quit India campaign; nonviolent protest can create change. Protest worked in the struggle for democracy in the Maldives. And on 24 October, we saw how protests across the world put Three-Five-O firmly on the Copenhagen agenda.

My message to you is to continue the protests. Continue after Copenhagen. Continue despite the odds. And eventually, together, we will reach that crucial number: Three-Five-O.

In all political agreements, you have to be prepared to negotiate. You have to be prepared to compromise; to give and take. That is the nature of politics. But physics isn't politics. On climate change, there are things on which we cannot negotiate. There are scientific bottom lines that we have to respect. We know what the laws of physics say. And I think you know too.

The most important number in the world. The most important number you'll ever hear. The most important number you'll ever say. These three words: Three-Five-O. Three-Five-O. Three-Five-O.

# PART III
# IMPACT

The End of Nature *was the first book for a general audience about global warming; it came out in 1989 and soon appeared in more than twenty languages. At the time, the available data on climate change still fit on the top of my desk. What's astonishing is that twenty years later most of the predictions scientists were then making have proved too conservative. Since climate change was still a new topic, I did my best to summarize the research on carbon and its effects, but I was more interested in how it made me feel—which was more sad than scared. Sad that we had ended the idea that any place was still beyond the human touch. Sad that our world no longer could claim to be wild.*

# from *The End of Nature*

Bill McKibben
*1989*

I took a day's hike last fall, walking Mill Creek from the spot where it runs by my door to the place where it crosses the mail county road near Wevertown. It's a distance of maybe nine miles as the car flies, but rivers are far less efficient, and endlessly follow pointless, time-wasting, uneconomical meanders and curves. Mill Creek cuts some fancy figures, and so I was able to feel a bit exploratory—a budget Bob Marshall. In a strict sense, it wasn't much of an adventure. I stopped at the store for a liverwurst sandwich at lunchtime, the path was generally downhill, the temperature stuck at an equable 55 degrees, and since it was the week before the hunting season opened I didn't have to sing as I walked to keep from getting shot. On the other hand, I had made an arbitrary plan—to follow the creek—and, as a consequence, I spent hours stumbling through overgrown marsh, batting at ten-foot saplings and vines, emerging only every now and then, scratched and weary, into the steeper wooded sections. When Thoreau was on Katahdin, nature said to him, "I have never made this soil for thy feet, this air for thy breathing, these rocks for thy neighbors. I cannot pity nor fondle thee there, but forever relentlessly drive thee hence to where I *am* kind. Why seek me where I have not called thee, and then complain because you find me but a stepmother?" Nature said this to me on Mill Creek, or at least it said, "Go home and tell your wife you walked to Wevertown." I felt I should have carried a machete, or employed a macheteist. (The worst thing about battling through brake and bramble of this sort is that it's so anonymous—gray sticks, green stalks with reddish thorns, none of them to be found in any of the many guides and almanacs on my shelf.) And though I started the day with eight dry socks, none saw noon in that pleasant state.

If it was all a little damp and in a minor key, the sky was nonetheless bright blue, and rabbits kept popping out from my path, and pheasants fired up between my legs, and at each turning some new gift appeared: a vein of quartz, or a ridge where the maples still held their leaves, or a pine more than three feet in diameter that beavers had gnawed all the way around and halfway through and then left standing—a forty-foot sculpture. It was October, so there weren't even any bugs. And always the plash of the stream in my ear. It isn't Yosemite, the Mill Creek Valley, but its small beauties are absorbing, and one can say with Muir on his mountaintop, "Up here all the world's prizes seem as nothing."

And so what if it isn't nature primeval? One of our neighbors has left several kitchen chairs along his stretch of the bank, spaced at fifty-yard intervals for comfort in fishing. At one old homestead, a stone chimney stands at either end of a foundation now filled by a graceful birch. Near the one real waterfall, a lot of rusty pipe and collapsed concrete testifies to the old mill that once stood there. But these aren't disturbing sights—they're almost comforting, reminders of the way that nature has endured and outlived and with dignity reclaimed so many schemes and disruptions of man. (A mile or so off the creek, there's a mine where a hundred and fifty years ago a visionary tried to extract pigment for paint and pack it out on mule and sledge. He rebuilt after a fire; finally an avalanche convinced him. The path in is faint now, but his chimney, too, still stands, a small Angkor Wat of free enterprise.) Large sections of the area were once farmed; but the growing season is not much more than a hundred days, and the limits established by that higher authority were stronger than the (powerful) attempts of individual men to circumvent them, and so the farms returned to forest, with only a dump of ancient bottles or a section of stone wall as a memorial. (Last fall, though, my wife and I found, in one abandoned meadow, a hop vine planted at least a century before. It was still flowering, and with its blossoms we brewed beer.) These ruins are humbling sights reminders of the negotiations with nature that have established the world as we know it.

Changing socks (soaking for merely clammy) in front of the waterfall, I thought back to the spring before last, when a record snowfall melted in only a dozen or so warm April days. A little to the south, an inflamed stream washed out a highway bridge, closing the New York Thruway for months. Mill Creek filled till it was a river, and this waterfall, normally one of those diaphanous-veil affairs, turned into a cataract. It filled me with awe to stand there then, on the shaking ground and think, This is what nature is capable of.

But as I sat there this time, and thought about the dry summer we'd just come through, there was nothing awe-inspiring or instructive, or even lulling, in the fall of the water. It suddenly seemed less like a waterfall than like a spillway to accommodate the overflow of a reservoir. That didn't decrease its beauty, but it changed its meaning. It has begun or will soon begin to rain and snow when the particular mix of chemicals we've injected into the atmosphere adds up to rain or snow—when they make it hot enough over some tropical sea to form a cloud and send it this way. I had no more control, in one sense, over this process than I ever did. But it felt different, and lonelier. Instead of a world where rain had an independent and mysterious existence, the rain had become a subset of human activity: a phenomenon like smog or commerce or the noise from the skidder towing logs on Cleveland Road—all things over which I had no control, either. The rain bore a brand; it was a steer, not a deer. And that was where the loneliness came from. There's nothing there except us. There's no such thing as nature anymore—that other world that isn't business and art and breakfast is now not another world, and there is nothing except us alone.

At the same time that I felt lonely, though, I also felt crowded, without privacy. We go to the woods in part to escape. But now there is nothing except us and so there is no escaping other people. As I walked in the autumn woods I saw a lot of sick trees. With the conifers, I suspected acid rain. (At least I have the luxury of only suspecting; in too many places, they *know*.) And so who walked with me in the woods? Well, there were the presidents of the Midwest utilities who kept explaining why they had to burn coal to make electricity (cheaper, fiduciary responsibility, no *proof* it kills trees) and then there were the congressmen who couldn't bring themselves to do anything about it (personally favor but politics the art of compromise, very busy with the war on drugs) and before long the whole human race had arrived to explain its aspirations. We like to drive, they said, air conditioning is a necessity nowadays, let's go to the mall. By this point, the woods were pretty densely populated. As I attempted to escape, I slipped on another rock, and in I went again. Of course, the person I was fleeing most fearfully was myself, for I drive (I drove forty thousand miles one year), and I'm burning a collapsed barn behind the house next week because it is much the cheapest way to deal with it, and I live on about four hundred times what Thoreau conclusively proved was enough, so I've done my share to take this independent, eternal world and turn it into a science-fair project (and not even a good science-fair project but a cloddish one, like pumping poison into an ant farm and "observing the effects").

The walk along Mill Creek, or any stream, or up any hill, or through any woods, is changed forever—changed as profoundly as when it shifted from pristine and untracked wilderness to mapped and deeded and cultivated land. Our local shopping mall now has a club of people who go "mall walking" every day. They circle the shopping center en masse—Caldor to Sears to J. C. Penney, circuit after circuit with an occasional break to shop. This seems less absurd to me now than it did at first. I like to walk in the outdoors not solely because the air is cleaner but because outdoors we venture into a sphere larger than ourselves. Mall walking involves too many other people, and too many purely human sights, ever to be more than good-natured exercise. But now, out in the wild, the sunshine on one's shoulders is a reminder that man has cracked the ozone, that, thanks to us, the atmosphere absorbs where once it released.

The greenhouse effect is a more apt name than those who coined it imagined. The carbon dioxide and trace gases act like the panes of glass on a greenhouse—the analogy is accurate. But it's more than that. We have built a greenhouse, a human creation, where once there bloomed a sweet and wild garden.

*One way to understand the impact of global warming is to think theologically, to hold up the current science against the religious understandings that have informed human beings throughout the course of civilization. This is a hard task: much of our theology reflects the historic idea that humans are small and God is powerful, a power evident in thunderstorms or the tossing sea. But now that ratio is shifting; increasingly it's led religious congregations to involve themselves in the effort to preserve creation. Sally Bingham is one of the leaders in that effort. With colleagues like Ben Webb and Steve McCausland, she founded the Regeneration Project and Episcopal Power and Light, now Interfaith Power and Light. With chapters across America, they've been an influential part of the struggle for climate legislation. Bingham gave this sermon at Stanford University in the spring of 2010.*

# John 5:1–9

Sally Bingham
*2010*

Thank you for having me this morning. It is a pleasure to be here at Stanford. I grew up in Woodside, just a few miles from here and it is very comforting to return to this area. I bring greeting from the Bishop of California, Marc Andrus. He recently installed me as Canon for the Environment and I believe he did that because he couldn't shut me up or keep me quiet about the destruction that humans have inflicted on creation. Around our diocese, I am referred to as the "loose canon"! I have no problem with that. I am a seeker of the truth and I intend to speak the truth even when the message is hard to hear. I am an environmentalist *because* I am a Christian, not in spite of it.

Today is Mother's Day. The Gospel lesson is about Jesus's ability to heal and my assignment is to talk about the environment. Can we bring all these things together? Yes!

Listen to what the former head of Yale's School of Forestry writes in the introduction to his latest book, *The Bridge at the Edge of the World*:

> Half the world's tropical and temperate forests are now gone. The rate of deforestation in the tropics continues

at about an acre a second. About half the wetlands and a third of the mangroves are gone. An estimated ninety percent of the large predator fish are gone, and 75 percent of marine fisheries are now over-fished or fished to capacity. Twenty percent of the corals are gone and another twenty percent severely threatened. Species are disappearing at rates about a thousand times faster than normal. The planet has not seen such a spasm of extinction in sixty-five million years, since the dinosaurs disappeared.

Ladies and gentlemen, we are standing at the edge of two worlds, the one that God created and the one that humans are making. The one that humans are making is not sustainable. It may be a matter of life and death. It is just that serious.

So let me ask you. Is this who we are? Was this God's purpose for the human species? I don't think so. We were given dominion over all that is. That is "dominion" not domination. It is dominion in the same sense that mothers have dominion over our children and that God has over us. Care and compassion, love and generosity. We don't give black eyes to our children or break their ribs; we don't purposely destroy their chance for a healthy life. Quite the contrary, we help our children in all ways possible to be the best they can be and that is what God expected from us when we were given dominion.

In light of Jesus's power to heal and our call to be his disciples, don't we have an obligation to do all that we can to prevent any harm to God's creation? In fact, in the Episcopal Church our Baptismal vows say, "I renounce any evil that destroys the creatures of God."

The new commandant that Jesus proclaimed in the gospel last week is paramount in this discussion: "Love one another!!

Everyone will know that you are my disciples, if you love one another." It could not be clearer. If you love your neighbor and one another, you don't pollute your neighbor's air or water and you certainly don't destroy your neighbor's right to a safe and healthy life. We cannot call that love, but rather action that is in direct disobedience to the commandment to love one another.

We are asked to love one another. That means not destroy the very basic stability that poor nations and poor people around the world need for survival. Yet it is the poor nations of the world (our neighbors) that are paying for mistakes that we in wealthy countries have made without their contribution or their knowledge. This is a justice situation and one that religions have come to see in just that way. Communities whose land, air and water are polluted cannot be healthy anymore than a young mother can nurture a child, if she is not healthy. We must try to keep young mothers healthy and keep our Mother Earth healthy. If those two things suffer from degradation the entire world suffers. Mothers are our source of life, our nourishment. Mother Earth is no different than any human mother. They must be healthy to support us and to produce healthy children.

People of faith are recognizing that destruction of creation is a moral issue and many religious leaders are getting involved and speaking out. Over the last ten years care for creation has taken hold in most major denominations. Religious leaders are protesting against blasting the tops of mountains and naming it a sin against God. Many congregations have formed green teams that are helping with reduction in energy use and less wasteful use of paper and plastic, caulking windows and replacing old appliances. These congregations are serving as examples to their community and showing individuals what they can do in their homes. A church in New Jersey had the congregation save its trash for two weeks and then open the black garbage bags and lay the contents all over the courtyard. The congregation was asked to put on rubber gloves and sort what was really garbage from what could be composted or recycled. It was a life-changing experience for many when they saw and touched what was going into landfill when it could have been reused, composted, or recycled.

While all this individual action is a step in the right direction, the destruction of creation is a global issue and needs to be addressed as such. Every one of our actions and behaviors affects someone else on the planet. Our individual choices matter, but we need every institution, every industry, every business, every industrial utility, every hospital, every school and every house of worship to be looking ahead at what we are leaving behind for our children and grandchildren to deal with and not taking risks with their health and safety.

Caring for God's creation takes on more urgency with each passing day. The climate continues to warm, half of all the earth's wetlands are already gone, we're losing an average of one species of plant or animal to extinction every hour, and all the while Americans are consuming their weight in stuff every day. This is *not* who we are, we don't have to do this—we have one earth and a shared purpose—to be the stewards of creation. We must keep our Mother Earth healthy.

I am often asked what the condition of the environment has to do with religion and if you are wondering that, too, it is a fair question, because for many of us the notion of environmental stewardship as a religious concern is relatively recent. But what's interesting is that throughout Scripture we are asked to love one another and to serve the poor. In our prayers we ask for the courage and will to use resources rightly without harm to anyone. Care for creation is a fundamental value of the Christian religion and also of other mainstream religions. Buddhists know and teach the interconnectedness of all things. If you harm any part you ultimately harm the whole. Muslims learn that God created the earth in balance and humans are called to keep that balance. Jews understand the lesson in Genesis that God put Adam in the garden to "till it and keep it."

You cannot sit in a pew and profess a love of God and let what God loves and called "good" be destroyed. God called the earth "good" before the Sierra Club existed.

There is a powerful reference that we hear every time we come to the altar. We proclaim it ourselves. In every Eucharistic prayer we proclaim the glory of God on earth. "Holy, holy, holy, Lord God Almighty, heaven and earth are full of your glory." The earth is full of God's glory. Listen anew to those words when you hear them this morning.

Besides Scripture there are other signs and signals that maybe we should pay more attention to. I have to raise this because it seems so poignant to me. There has been a lot of discussion lately about opening up our coastlands to more offshore drilling and no sooner has it come into public debate than we have an explosion, eleven people die tragically in the Gulf of Mexico and miles and miles of coastland are ruined. This oil disaster will cost billions of dollars to clean up and much of the wildlife will never recover. There is pristine wilderness and many wildlife refuge areas that will be ruined forever. Louisiana is only beginning to recover from Katrina and now this disaster hits them. But it affects us all. An oil disaster that will have negative effects for years to come. Are we listening to the signs around us? Do our ears hear and our eyes see?

Could we do it differently?

There are many groups who think that coal mining is no longer viable. A dirty business, a thing of the past—yet we keep sending miners down into mile long tunnels to bring it up because it is cheap. This subject, too, is under hot debate when a tragic mine disaster occurred. Lives are lost and land is destroyed forever. Are we listening to the signs? I wonder. What will it take for us to be willing to change our ways? When might we measure the true cost of both offshore drilling and coal mining?

For two hundred years we have been burning fossil fuel for energy and in the last twenty years, scientists have discovered that this form of energy is harming creation. Greenhouse gases, as they are called, have formed a blanketlike cover that is preventing heat from escaping and thus warming the planet. The fuels we are heavily dependent upon are finite, dirty, and dangerous. The time as come to switch to clean renewable energy that comes from sun and wind, geothermal and other resources that don't pollute the air with carbon dioxide and other harmful gases. Why now? Well, the situation has reached a crisis to the extent that many of our most influential scientists think we may already have reached a point of no return.

What does Jesus say about change or switching to new ways and ideas? Remember when he told the disciples to put their nets on the other side of the boat. He says, "If you're not catching any fish on one side of the boat—try the other side." He says, Think out of the box, or in this case you could say "Think out of the barrel." There is more than one way to do something. Open your minds to a new way of doing things. I cannot possibly suggest what Jesus would do if he were aware of what humans have done to the planet, but I suspect that, being a healer, he would want to help the poor nations of the world that are disappearing because of sea rise. I suspect that he would want to save endangered species of all kinds, restore forests and wetlands to health, prevent the ice caps from melting. I think he would say, Put your nets on the other side of the boat. What you are doing isn't working; try something else. There are abundant resources; you just have to think creatively, in a new way.

Jesus continually used natural images in his teaching and while they were simple images like fig trees bearing fruit or mustard seeds or fishing from the other side of the boat, these teachings have profound meaning for us. Fishing from the other side of the boat is such a strong metaphor for *do it differently* when the old way isn't working. If being dependent on fossil fuel for energy isn't working, try something else. (The fuels from heaven might be an alternative. Sun and wind.)

Remember the lesson in which Jesus commanded Peter to "feed my sheep"? Certainly, this feeding refers to the spiritual food of the Gospel and the Eucharist. But it can also extend to the literal, physical sense that we must feed our sisters and brothers who are hungry. Another call to serve one another, and in this case do all in our power to ensure that people around the world are not starving due to our behavior. If global warming is extending droughts in certain parts of Australia and Africa, as science says it is, and people are unable to find or grow food, don't we have a responsibility to help out?

What I suggest for us is a new way of thinking about the human purpose on earth. Why are we here and, in light of the climate crisis and other destructive measures that humans have visited on God's Creation, don't we have a shared responsibility?

Is this our shared purpose? Crisis can and often does mean danger, but it also is an opportunity, and we mustn't waste this opportunity. We have the chance to do what few generations have had, and that is to come together with a shared purpose and the joy of working together as a unified force with a regenerated view of the human purpose on earth. Let's accept the view that we are one—one earth, one atmosphere, and one global climate with one shared purpose—to look after and love one another: to do our best to heal what is broken or not working, and to treat Mother Earth with love. "By this everyone will know that you are my disciples, if you have love for one another." Amen.

*For many years, more conservative Christian churches were wary of involvement with environmentalism; many evangelicals and fundamentalists considered ecology akin to paganism. But recently, recognizing the threat climate change poses to creation, involvement has grown, especially among younger congregants. In 2006, after long consultation with scientists, eighty-six prominent evangelical leaders (including Rick Warren, the author of the mega-bestseller* The Purpose Driven Life*) and the heads of many evangelical seminaries issued the following letter. It was one of the first public breaks between evangelicals and the Republican Party, and it drew strong reaction from some right-wing Christian leaders.*

# Climate Change: An Evangelical Call to Action

The Evangelical Climate Initiative
*2006*

*Preamble*

As American evangelical Christian leaders, we recognize both our opportunity and our responsibility to offer a biblically based moral witness that can help shape public policy in the most powerful nation on earth, and therefore contribute to the well-being of the entire world.[1] Whether we will enter the public square and offer our witness there is no longer an open question. We are in that square, and we will not withdraw.

We are proud of the evangelical community's long-standing commitment to the sanctity of human life. But we also offer moral witness in many venues and on many issues. Sometimes the issues that we have taken on, such as sex trafficking, genocide in the Sudan, and the AIDS epidemic in Africa, have surprised outside observers. While individuals and organizations can be called to concentrate on certain issues, we are not a single-issue movement. We seek to be true to our calling as Christian leaders, and above all faithful to Jesus Christ our Lord. Our attention, therefore, goes to whatever issues our faith requires us to address.

Over the last several years many of us have engaged in study, reflection, and prayer related to the issue of climate change (often called "global warming"). For most of us, until recently this has not been treated as a pressing issue or major priority. Indeed, many of us have required considerable convincing before becoming persuaded that climate change is a real problem and that it ought to matter to us as Christians. But now we have seen and heard enough to offer the following moral argument related to the matter of human-induced climate change. We commend the four simple but urgent claims offered in this document to all who will listen, beginning with our brothers and sisters in the Christian community, and urge all to take the appropriate actions that follow from them.

## Claim 1: Human-Induced Climate Change is Real

Since 1995 there has been general agreement among those in the scientific community most seriously engaged with this issue that climate change is happening and is being caused mainly by human activities, especially the burning of fossil fuels. Evidence gathered since 1995 has only strengthened this conclusion.

Because all religious/moral claims about climate change are relevant only if climate change is real and is mainly human-induced, everything hinges on the scientific data. As evangelicals we have hesitated to speak on this issue until we could be more certain of the science of climate change, but the signatories now believe that the evidence demands action:

The Intergovernmental Panel on Climate Change (IPCC), the world's most authoritative body of scientists and policy experts on the issue of global warming, has been studying this issue since the late 1980s. (From 1988 to 2002 the IPCC's assessment of the climate science was Chaired by Sir John Houghton, a devout evangelical Christian.) It has documented the steady rise in global temperatures over the last fifty years, projects that the average global temperature will continue to rise in the coming decades, and attributes "most of the warming" to human activities.

The U.S. National Academy of Sciences, as well as all other G8 countries' scientific academies (Great Britain, France, Germany, Japan, Canada, Italy, and Russia), has concurred with these judgments.

In a 2004 report, and at the 2005 G8 summit, the Bush Administration has also acknowledged the reality of climate change and the likelihood that human activity is the cause of at least some of it.[2]

In the face of the breadth and depth of this scientific and governmental concern, only a small percentage of which is noted here, we are convinced that evangelicals must engage this issue without any further lingering over the basic reality of the problem or humanity's responsibility to address it.

### Claim 2: The Consequences of Climate Change Will Be Significant, and Will Hit the Poor the Hardest

The earth's natural systems are resilient but not infinitely so, and human civilizations are remarkably dependent on ecological stability and well-being. It is easy to forget this until that stability and well-being are threatened.

Even small rises in global temperatures will have such likely impacts as: sea level rise; more frequent heat waves, droughts, and extreme weather events such as torrential rains and floods; increased tropical diseases in now-temperate regions; and hurricanes that are more intense. It could lead to significant reduction in agricultural output, especially in poor countries. Low-lying regions, indeed entire islands, could find themselves under water.

(This is not to mention the various negative impacts climate change could have on God's other creatures.)

Each of these impacts increases the likelihood of refugees from flooding or famine, violent conflicts, and international instability, which could lead to more security threats to our nation.

Poor nations and poor individuals have fewer resources available to cope with major challenges and threats. The consequences of global warming will therefore hit the poor the hardest, in part because those areas likely to be significantly affected first are in the poorest regions of the world. Millions of people could die in this century because of climate change, most of them our poorest global neighbors.

*Claim 3: Christian Moral Convictions Demand Our Response to the Climate Change Problem*

While we cannot here review the full range of relevant biblical convictions related to care of the creation, we emphasize the following points:

Christians must care about climate change because we love God the Creator and Jesus our Lord, through whom and for whom the creation was made. This is God's world, and any damage that we do to God's world is an offense against God Himself (Genesis 1; Psalms 24; Colossians 1:16).

Christians must care about climate change because we are called to love our neighbors, to do unto others as we would have them do unto us, and to protect and care for the least of these as though each was Jesus Christ himself (Matthew 22:34–40; Matthew 7:12; Matthew 25:31–46).

Christians, noting the fact that most of the climate change problem is human-induced, are reminded that when God made humanity he commissioned us to exercise stewardship over the earth and its creatures. Climate change is the latest evidence of our failure to exercise proper stewardship, and constitutes a critical opportunity for us to do better (Genesis 1:26–28).

Love of God, love of neighbor, and the demands of stewardship are more than enough reason for evangelical Christians to respond to the climate change problem with moral passion and concrete action.

*Claim 4: The need to act now is urgent. Governments, businesses, churches, and individuals all have a role to play in addressing climate change—starting now.*

The basic task for all of the world's inhabitants is to find ways now to begin to reduce the carbon dioxide emissions from the burning of fossil fuels that are the primary cause of human-induced climate change.

There are several reasons for urgency. First, deadly impacts are being experienced now. Second, the oceans only warm slowly, creating a lag in experiencing the consequences. Much of the climate change to which we are already committed will not be realized for several decades. The consequences of the pollution we create now will be visited upon our children and grandchildren. Third, as individuals and as a society we are making long-term decisions today that will determine how much carbon dioxide we will emit in the future, such as whether to purchase energy efficient vehicles and appliances that will last for ten to twenty years, or whether to build more coal-burning power plants that last for fifty years rather than investing more in energy efficiency and renewable energy.

In the United States, the most important immediate step that can be taken at the federal level is to pass and implement national legislation requiring sufficient economy-wide reductions in carbon dioxide emissions through cost-effective, market-based mechanisms such as a cap-and-trade program. On June 22, 2005 the Senate passed the Domenici-Bingaman resolution affirming this approach, and a number of major energy companies now acknowledge that this method is best both for the environment and for business.

We commend the Senators who have taken this stand and encourage them to fulfill their pledge. We also applaud the steps taken by such companies as BP, Shell, General Electric, Cinergy, Duke Energy, and DuPont, all of which have moved ahead of the pace of government action through innovative measures implemented within their companies in the U.S. and around the world. In so doing they have offered timely leadership.

Numerous positive actions to prevent and mitigate climate change are being implemented across our society by state and local governments, churches, smaller businesses, and individuals. These commendable efforts focus on such matters as energy efficiency, the use of renewable energy, low $CO_2$ emitting technologies, and the purchase of hybrid vehicles. These efforts can easily be shown to save money, save energy, reduce global warming pollution as well as air pollution that harm human health, and eventually pay for themselves. There is much more to be done, but these pioneers are already helping to show the way forward.

Finally, while we must reduce our global warming pollution to help mitigate the impacts of climate change, as a society and as individuals we must also help the poor adapt to the significant harm that global warming will cause.

# Conclusion

We the undersigned pledge to act on the basis of the claims made in this document. We will not only teach the truths communicated here but also seek ways to implement the actions that follow from them. In the name of Jesus Christ our Lord, we urge all who read this declaration to join us in this effort.

## Signatories

Institutional affiliation is given for identification purposes only. All signatories do so as individuals expressing their personal opinions and not as representatives of their organizations.

Rev. Dr. Leith Anderson, Former President, National Association of Evangelicals (NAE); Senior Pastor, Wooddale Church; Eden Prairie, MN

Robert Andringa, Ph.D., President, Council for Christian Colleges and Universities (CCCU); Vienna, VA

Rev. Jim Ball, Ph.D., Executive Director, Evangelical Environmental Network; Wynnewood, PA

Commissioner W. Todd Bassett, National Commander, The Salvation Army; Alexandria, VA

Dr. Jay A. Barber, Jr., President, Warner Pacific College; Portland, OR

Gary P. Bergel, President, Intercessors for America; Purcellville, VA

David Black, Ph.D., President, Eastern University; St. Davids, PA

Bishop Charles E. Blake, Sr., West Angeles Church of God in Christ, Los Angeles, CA

Rev. Dr. Dan Boone, President, Trevecca Nazarene University; Nashville, TN

Bishop Wellington Boone, The Father's House & Wellington Boone Ministries; Norcross, GA

Rev. Dr. Peter Borgdorff, Executive Director, Christian Reformed Church; Grand Rapids, MI

H. David Brandt, Ph.D., President, George Fox University; Newberg, OR

Rev. George K. Brushaber, Ph.D., President, Bethel University; Senior Advisor, *Christianity Today*; St. Paul, MN

Rev. Dwight Burchett, President, Northern California Association of Evangelicals; Sacramento, CA

Gaylen Byker, Ph.D., President, Calvin College; Grand Rapids, MI

Rev. Dr. Jerry B. Cain, President, Judson College; Elgin, IL

Rev. Dr. Clive Calver, Senior Pastor, Walnut Hill Community Church; Former President, World Relief; Bethel, CT

R. Judson Carlberg, Ph.D., President, Gordon College; Wenham, MA

Rev. Dr. Paul Cedar, Chair, Mission America Coalition; Palm Desert, CA

David Clark, Ph.D., President, Palm Beach Atlantic University; Former Chair/CEO, Nat. Rel. Broadcasters; Founding Dean, Regent University; West Palm Beach, FL

Rev. Luis Cortes, President & CEO, Esperanza USA; Host, National Hispanic Prayer Breakfast; Philadelphia, PA

Andy Crouch, Columnist, *Christianity Today* magazine; Swarthmore, PA

Rev. Paul de Vries, Ph.D., President, New York Divinity School; New York, NY

Rev. David S. Dockery, Ph.D., Chairman of the Board, Council for Christian Colleges and Universities; President, Union University; Jackson, TN

Larry R. Donnithorne, Ed.D., President, Colorado Christian University; Lakewood, CO

Blair Dowden, Ed.D., President, Huntington University; Huntington, IN

Rev. Robert P. Dugan, Jr., Former VP of Governmental Affairs, National Association of Evangelicals; Palm Desert, CA

Craig Hilton Dyer, President, Bright Hope International, Hoffman Estates, IL D.

Merrill Ewert, Ed.D., President, Fresno Pacific University; Fresno, CA

Rev. Dr. LeBron Fairbanks, President, Mount Vernon Nazarene University; Mount Vernon, OH

Rev. Myles Fish, President/CEO, International Aid, Spring Lake, MI

Rev. Dr. Floyd Flake, Senior Pastor, Greater Allen AME Cathedral; President, Wilberforce University; Jamaica, NY

Rev. Timothy George, Ph.D., Founding Dean, Beeson Divinity School, Samford University, Executive Editor, *Christianity Today*; Birmingham, AL

Rev. Michael J. Glodo, Stated Clerk, Evangelical Presbyterian Church; Livonia, MI

Rev. James M. Grant, Ph.D., President, Simpson University; Redding, CA

Rev. Dr. Jeffrey E. Greenway, President, Asbury Theological Seminary; Wilmore, KY

Rev. David Gushee, Professor of Moral Philosophy, Union University; columnist, Religion News Service; Jackson, TN

Gregory V. Hall, President, Warner Southern College; Lake Wales, FL

Brent Hample, Executive Director, India Partners; Eugene, OR

Rev. Dr. Jack Hayford, President, International Church of the Foursquare Gospel; Los Angeles, CA

Rev. Steve Hayner, Ph.D., Former President, InterVarsity; Prof. of Evangelism, Columbia Theological Sem.; Decatur, GA

E. Douglas Hodo, Ph.D., President, Houston Baptist University; Houston, TX

Ben Homan, President, Food for the Hungry; President, Association of Evangelical Relief and Development Organizations (AERDO); Phoenix, AZ

Rev. Dr. Joel Hunter, Senior Pastor, Northland, A Church Distributed; Longwood, FL

Bryce Jessup, President; William Jessup University, Rocklin, CA

Ronald G. Johnson, Ph.D., President, Malone College; Canton, OH

Rev. Dr. Phillip Charles Joubert, Sr., Pastor, Community Baptist Church; Bayside, NY

Jennifer Jukanovich, Founder, The Vine; Seattle, WA

Rev. Brian Kluth, Senior Pastor, First Evangelical Free Church; Founder, MAXIMUM Generosity; Colorado Springs, CO

Bishop James D. Leggett, General Superintendent, International Pentecostal Holiness Church; Chair, Pentecostal World Fellowship; Oklahoma City, OK

Duane Litfin, Ph.D., President, Wheaton College; Wheaton, IL

Rev. Dr. Larry Lloyd, President, Crichton College; Memphis, TN

Rev. Dr. Jo Anne Lyon, Executive Director, World Hope; Alexandria, VA

Sammy Mah, President and CEO, World Relief; Baltimore, MD

Jim Mannoia, Ph.D., President, Greenville College; Greenville, IL

Bishop George D. McKinney, Ph.D., D.D., St. Stephens Church Of God In Christ; San Diego, CA

Rev. Brian McLaren, Senior Pastor, Cedar Ridge Community Church; Emergent leader; Spencerville, MD

Rev. Dr. Daniel Mercaldo, Senior Pastor & Founder, Gateway Cathedral; Staten Island, NY

Rev. Dr. Jesse Miranda, President, AMEN; Costa Mesa, CA

Royce Money, Ph.D., President, Abilene Christian University; Abilene, TX

Dr. Bruce Murphy, President, Northwestern University; Orange City, IA

Rev. George W. Murray, D.Miss., President, Columbia International University; Columbia SC

David Neff, Editor, *Christianity Today*; Carol Stream, IL

Larry Nikkel, President, Tabor College; Hillsboro, KS

Michael Nyenhuis, President, MAP International; Brunswick, GA

Brian O'Connell, President, REACT Services; Founder and Former Executive Director, Religious Liberty Commission, World Evangelical Alliance; Mill Creek, WA

Roger Parrott, Ph.D., President, Belhaven College; Jackson, MS

Charles W. Pollard, Ph.D., J.D., President, John Brown University; Siloam Springs, AR

Paul A. Rader, D.Miss., President, Asbury College; Wilmore, KY

Rev. Edwin H. Robinson, Ph.D., President, MidAmerica Nazarene Univ., Olathe, KS

William P. Robinson, Ph.D., President, Whitworth College; Spokane, WA

Lee Royce, Ph.D., President, Mississippi College; Clinton, MS

Andy Ryskamp, Executive Director, Christian Reformed World Relief Committee; Grand Rapids, MI

Rev. Ron Sider, Ph.D., President, Evangelicals for Social Action; Philadelphia, PA

Richard Stearns, President, World Vision; Federal Way, WA

Rev. Jewelle Stewart, Ex. Dir., Women's Ministries, International Pentecostal Holiness Church; Oklahoma City, OK

Rev. Dr. Loren Swartzendruber, President, Eastern Mennonite University; Harrisonburg, VA

C. Pat Taylor, Ph.D., President, Southwest Baptist University; Bolivar, MO

Rev. Berten A. Waggoner, National Director, Vineyard, USA; Sugar Land, TX

Jon R. Wallace, DBA, President, Azusa Pacific University; Azusa, CA

Rev. Dr. Thomas Yung-Hsin Wang, former International Director of Lausanne II; Sunnyvale, CA

Rev. Dr. Rick Warren, Senior Pastor, Saddleback Church; author of *The Purpose Driven Life*; Lake Forest, CA

John Warton, President, Business Professional Network; Portland, OR

Robert W. Yarbrough, Ph.D., New Testament Dept. Chair, Trinity Evangelical Divinity School; Deerfield, IL

John D. Yordy, Ph.D., Interim President, Goshen College; Goshen, IN

Adm. Tim Ziemer, Director of Programs, World Relief; Baltimore, MD

*One lens through which to regard the prospect of rapid climate change is global security: as events like 2010's Pakistan floods made clear, it will be hard to maintain peace in a world undergoing the kinds of stresses that now seem almost inevitable. In the winter of 2004, a pair of "futurists" prepared this report for an arm of the Pentagon tasked with anticipating future challenges. Many of its particular predictions, including the prospect of a shutdown of the Gulf Stream, seem unlikely according to current science. But the simple notion that even George W. Bush's Defense Department was taking global warming seriously helped shape public opinion.*

# An Abrupt Climate Change Scenario and Its Implications for United States National Security

Peter Schwartz and Doug Randall
*2003*

## *Imagining the Unthinkable*

The purpose of this report is to imagine the unthinkable—to push the boundaries of current research on climate change so we may better understand the potential implications on United States national security.

We have interviewed leading climate change scientists, conducted additional research, and reviewed several iterations of the scenario with these experts. The scientists support this project, but caution that the scenario depicted is extreme in two fundamental ways. First, they suggest the occurrences we outline would most likely happen in a few regions, rather than on globally. Second, they say the magnitude of the event may be considerably smaller.

We have created a climate change scenario that although not the most likely, is plausible, and would challenge United States national security in ways that should be considered immediately.

## *Executive Summary*

There is substantial evidence to indicate that significant global warming will occur during the twenty-first century. Because changes have been gradual so far, and are projected to be similarly gradual in the future, the effects of global warming have the potential to be manageable for most nations. Recent research, however, suggests that there is a possibility that this gradual global warming could lead to a relatively abrupt slowing of the ocean's thermohaline conveyor, which could lead to harsher winter weather conditions, sharply reduced soil moisture, and more intense winds in certain regions that currently provide a significant fraction of the world's food production. With inadequate preparation, the result could be a significant drop in the human carrying capacity of the Earth's environment.

The research suggests that once temperature rises above some threshold, adverse weather conditions could develop relatively abruptly, with persistent changes in the atmospheric circulation causing drops in some regions of five to ten degrees Fahrenheit in a single decade. Paleoclimatic evidence suggests that altered climatic patterns could last for as much as a century, as they did when the ocean conveyor collapsed 8,200 years ago, or, at the extreme, could last as long as 1,000 years as they did during the Younger Dryas, which began about 12,700 years ago.

In this report, as an alternative to the scenarios of gradual climatic warming that are so common, we outline an abrupt climate change scenario patterned after the 100-year event that occurred about 8,200 years ago. This abrupt change scenario is characterized by the following conditions:

- Annual average temperatures drop by up to five degrees Fahrenheit over Asia and North America and six degrees Fahrenheit in northern Europe.
- Annual average temperatures increase by up to four degrees Fahrenheit in key areas throughout Australia, South America, and southern Africa.

- Drought persists for most of the decade in critical agricultural regions and in the water resource regions for major population centers in Europe and eastern North America.
- Winter storms and winds intensify, amplifying the impacts of the changes. Western Europe and the North Pacific experience enhanced winds.

The report explores how such an abrupt climate change scenario could potentially destabilize the geo-political environment, leading to skirmishes, battles, and even war due to resource constraints such as:

1. Food shortages due to decreases in net global agricultural production
2. Decreased availability and quality of fresh water in key regions due to shifted precipitation patters, causing more frequent floods and droughts
3. Disrupted access to energy supplies due to extensive sea ice and storminess

As global and local carrying capacities are reduced, tensions could mount around the world, leading to two fundamental strategies: defensive and offensive. Nations with the resources to do so may build virtual fortresses around their countries, preserving resources for themselves. Less fortunate nations especially those with ancient enmities with their neighbors, may initiate struggles for access to food, clean water, or energy. Unlikely alliances could be formed as defense priorities shift and the goal is resources for survival rather than religion, ideology, or national honor.

This scenario poses new challenges for the United States, and suggests several steps to be taken:

- Improve predictive climate models to allow investigation of a wider range of scenarios and to anticipate how and where changes could occur.
- Assemble comprehensive predictive models of the potential impacts of abrupt climate change to improve projections of how climate could influence food, water, and energy.
- Create vulnerability metrics to anticipate which countries are most vulnerable to climate change and therefore, could contribute materially to an increasingly disorderly and potentially violent world.
- Identify no-regrets strategies such as enhancing capabilities for water management.
- Rehearse adaptive responses.
- Explore local implications.
- Explore geo-engineering options that control the climate.

There are some indications today that global warming has reached the threshold where the thermohaline circulation could start to be significantly impacted. These indications include observations documenting that the North Atlantic is increasingly being freshened by melting glaciers, increased precipitation, and fresh water runoff making it substantially less salty over the past forty years.

This report suggests that, because of the potentially dire consequences, the risk of abrupt climate change, although uncertain and quite possibly small, should be elevated beyond a scientific debate to a U.S. national security concern.

## Introduction

When most people think about climate change, they imagine gradual increases in temperature and only marginal changes in other climatic conditions, continuing indefinitely or even leveling off at some time in the future. The conventional wisdom is that modern civilization will either adapt to whatever weather conditions we face and that the pace of climate change will not overwhelm the adaptive capacity of society, or that our efforts such as those embodied in the Kyoto protocol will be sufficient to mitigate the impacts. The IPCC documents the threat of gradual climate change and its impact to food supplies and other resources of importance to humans will not be so severe as to create security threats. Optimists assert that the benefits from technological innovation will be able to outpace the negative effects of climate change.

Climatically, the gradual-change view of the future assumes that agriculture will continue to thrive and growing seasons will lengthen. Northern Europe, Russia, and North America will prosper agriculturally while southern Europe, Africa, and Central and South America will suffer from increased dryness, heat, water shortages, and reduced production. Overall, global food production under many typical climate scenarios increases. This view of climate change may be a dangerous act of self-deception, as increasingly we are facing weather related disasters—more hurricanes, monsoons, floods, and dry spells—in regions around the world.

Weather-related events have an enormous impact on society, as they influence food supply, conditions in cities and communities, as well as access to clean water and energy. For example, a recent report by the Climate Action Network of Australia projects that climate change is likely to reduce rainfall in the rangelands, which could lead to a fifteen-percent drop in grass productivity. This, in turn, could lead to reductions in the average weight of cattle by twelve percent, significantly reducing beef supply.

Under such conditions, dairy cows are projected to produce thirty percent less milk, and new pests are likely to spread in fruit-growing areas. Additionally, such conditions are projected to lead to ten percent less water for drinking. Based on model projections, change conditions such as these could occur in several food-producing regions around the world at the same time within the next fifteen to thirty years, challenging the notion that society's ability to adapt will make climate change manageable.

With over 400 million people living in drier, subtropical, often over-populated and economically poor regions today, climate change and its follow-on effects pose a severe risk to political, economic, and social stability. In less prosperous regions, where countries lack the resources and capabilities required to adapt quickly to more severe conditions, the problem is very likely to be exacerbated. For some countries, climate change could become such a challenge that mass emigration results as the desperate peoples seek better lives in regions such as the United States that have the resources to adaptation.

Because the prevailing scenarios of gradual global warming could cause effects like the ones described above, an increasing number of business leaders, economists, policy makers, and politicians are concerned about the projections for further change and are working to limit human influences on the climate. But these efforts may not be sufficient or be implemented soon enough.

Rather than decades or even centuries of gradual warming, recent evidence suggests the possibility that a more dire climate scenario may actually be unfolding. This is why GBN [the Global Business Network] is working with OSD [the Office of the Secretary of Defense] to develop a plausible scenario for abrupt climate change that can be used to explore implications for food supply, health and disease, commerce and trade, and their consequences for national security.

While future weather patterns and the specific details of abrupt climate change cannot be predicted accurately or with great assurance, the actual history of climate change provides some useful guides. Our goal is merely to portray a plausible scenario, similar to one which has already occurred in human experience, for which there is reasonable evidence so that we may further explore potential implications for United States national security.

CREATING THE SCENARIO: REVIEWING HISTORY

## The Cooling Event 8,200 Years Ago

The climate change scenario outlined in this report is modeled on a century-long climate event that records from an ice core in Greenland indicate occurred 8,200 years ago. Immediately following an extended period of warming, much like the phase we appear to be in today, there was a sudden cooling. Average annual temperatures in Greenland dropped by roughly five degrees Fahrenheit, and temperature decreases nearly this large are likely to have occurred throughout the North Atlantic region. During the 8,200 event, severe winters in Europe and some other areas caused glaciers to advance, rivers to freeze, and agricultural lands to be less productive. Scientific evidence suggests that this event was associated with, and perhaps caused by, a collapse of the ocean's conveyor following a period of gradual warming.

Longer ice-core and oceanic records suggest that there may have been as many as eight rapid cooling episodes in the past

730,000 years, and sharp reductions in the ocean conveyer—a phenomenon that may well be on the horizon—are a likely suspect in causing such shifts in climate.

## The Younger Dryas

About 12,700 years ago, also associated with an apparent collapse of the thermohaline circulation, there was a cooling of at least twenty-seven degrees Fahrenheit in Greenland, and substantial change throughout the North Atlantic region as well, this time lasting 1,300 years. The remarkable feature of the Younger Dryas event was that it happened in a series of decadal drops of around five degrees, and then the cold, dry weather persisted for over 1,000 years. While this event had an enormous effect on the ocean and land surrounding Europe (causing icebergs to be found as far south as the coast of Portugal), its impact would be more severe today—in our densely populated society. It is the more recent periods of cooling that appear to be intimately connected with changes to civilization, unrest, inhabitability of once-desirable land, and even the demise of certain populations.

## The Little Ice Age

Beginning in the fourteenth century, the North Atlantic region experienced a cooling that lasted until the mid-nineteenth century. This cooling may have been caused by a significant slowing of the ocean conveyor, although it is more generally thought that reduced solar output and/or volcanic eruptions may have prompted the oceanic changes. This period, often referred to as the Little Ice Age, which lasted from 1300 to 1850, brought severe winters, sudden climatic shifts, and profound agricultural, economic, and political impacts to Europe.

The period was marked by persistent crop failures, famine, disease, and population migration, perhaps most dramatically felt by the Norse, also known as the Vikings, who inhabited Iceland and later Greenland. Ice formations along the coast of Greenland prevented merchants from getting their boats to Greenland and fisherman from getting fish for entire winters. As a result, farmers were forced to slaughter their poorly fed livestock—because of a lack of food both for the animals and themselves—but without fish, vegetables, and grains, there was not enough food to feed the population.

Famine, caused in part by the more severe climatic conditions, is reported to have caused tens of thousands of deaths between 1315 and 1319 alone. The general cooling also apparently drove the Vikings out of Greenland—and some say was a contributing cause for that society's demise.

While climate crises like the Little Ice Age aren't solely responsible for the death of civilizations, it's undeniable that they have a large impact on society. It has been less than 175 years since one million people died due to the Irish Potato Famine, which also was induced in part by climate change.

## A Climate Change Scenario For the Future

The past examples of abrupt climate change suggest that it is prudent to consider an abrupt climate change scenario for the future as plausible, especially because some recent scientific findings suggest that we could be on the cusp of such an event. The future scenario that we have constructed is based on the 8,200 years before present event, which was much warmer and far briefer than the Younger Dryas, but more severe than the Little Ice Age. This scenario makes plausible assumptions about which parts of the globe are likely to be colder, drier, and windier. Although intensified research could help to refine the assumptions, there is no way to confirm the assumptions on the basis of present models.

Rather than predicting how climate change will happen, our intent is to dramatize the impact climate change could have on society if we are unprepared for it. Where we describe concrete weather conditions and implications, our aim is to further the strategic conversation rather than to accurately forecast what is likely to happen with a high degree of certainty. Even the most sophisticated models cannot predict the details of how the climate change will unfold, which regions will be impacted in which ways, and how governments and society might respond. However, there appears to be general agreement in the scientific community that an extreme case like the one depicted below is not implausible. Many scientists would regard this scenario as extreme both in how soon it develops, and how large, rapid and ubiquitous the climate changes are. But history tells us that sometimes the extreme cases do occur, there is evidence that it might be, and it is DOD's [the Department of Defense's] job to consider such scenarios.

Keep in mind that the duration of this event could be decades, centuries, or millennia, and it could begin this year or many years in the future. In the climate change disruption scenario proposed here, we consider a period of gradual warming leading to 2010 and then outline the following ten years, when, like in the 8,200 event, an abrupt change toward cooling in the pattern of weather conditions change is assumed to occur.

## Warming Up to 2010

Following the most rapid century of warming experienced by modern civilization, the first ten years of the twenty-first century see an acceleration of atmospheric warming, as average temperatures worldwide rise by 0.5 degrees Fahrenheit per decade and by as much as two degrees Fahrenheit per decade in the harder-hit regions. Such temperature changes would vary both by region and by season over the globe, with these finer-scale variations being larger or smaller than the average change. What would be very clear is that the planet is continuing the warming trend of the late twentieth century.

Most of North America, Europe, and parts of South America experience thirty percent more days with peak temperatures over ninety degrees Fahrenheit than they did a century ago, with far fewer days below freezing. In addition to the warming, there are erratic weather patterns: more floods, particularly in mountainous regions, and prolonged droughts in grain-producing and coastal-agricultural areas. In general, the climate shift is an economic nuisance, generally affecting local areas as storms, droughts, and hot spells impact agriculture and other climate-dependent activities. (More French doctors remain on duty in August, for example.) The weather pattern, though, is not yet severe enough or widespread enough to threaten the interconnected global society or United States national security.

## Warming Feedback Loops

As temperatures rise throughout the twentieth century and into the early 2000s, potent positive feedback loops kick in, accelerating the warming from 0.2 degrees Fahrenheit, to 0.4 and eventually 0.5 degrees Fahrenheit per year in some locations. As the surface warms, the hydrologic cycle (evaporation, precipitation, and runoff) accelerates, causing temperatures to rise even higher. Water vapor, the most powerful natural greenhouse gas, traps additional heat and brings average surface air temperatures up. As evaporation increases, higher surface air temperatures cause drying in forests and grasslands, where animals graze and farmers grow grain. As trees die and burn, forests absorb less carbon dioxide, again leading to higher surface air temperatures as well as fierce and uncontrollable forest fires. Further, warmer temperatures melt snow cover in mountains, open fields, high-latitude tundra areas, and permafrost throughout forests in cold-weather areas. With the ground absorbing more and reflecting less of the sun's rays, temperatures increase even higher.

By 2005, the climatic impact of the shift is felt more intensely in certain regions around the world. More severe storms and typhoons bring about higher storm surges and floods in low-lying islands such as Tarawa and Tuvalu (near New Zealand). In 2007, a particularly severe storm causes the ocean to break through levees in the Netherlands making a few key coastal cities such as The Hague unlivable. Failures of the delta island levees in the Sacramento River region in the Central Valley of California creates an inland sea and disrupts the aqueduct system transporting water from northern to southern California because salt water can no longer be kept out of the area during the dry season. Melting along the Himalayan glaciers accelerates, causing some Tibetan people to relocate. Floating ice in the northern polar seas, which had already lost forty percent of its mass from 1970 to 2003, is mostly gone during summer by 2010. As glacial ice melts, sea levels rise and as wintertime sea extent decreases, ocean waves increase in intensity, damaging coastal cities. Additionally millions of people are put at risk of flooding around the globe (roughly four times 2003 levels), and fisheries are disrupted as water temperature changes cause fish to migrate to new locations and habitats, increasing tensions over fishing rights.

Each of these local disasters caused by severe weather impacts surrounding areas whose natural, human, and economic resources are tapped to aid in recovery. The positive feedback loops and acceleration of the warming pattern begin to trigger responses that weren't previously imagined, as natural disasters and stormy weather occur in both developed and lesser-developed nations. Their impacts are greatest in less-resilient developing nations, which do not have the capacity built into their social, economic, and agricultural systems to absorb change.

As melting of the Greenland ice sheet exceeds the annual snowfall, and there is increasing freshwater runoff from high latitude precipitation, the freshening of waters in the North Atlantic Ocean and the seas between Greenland and Europe increases. The lower densities of these freshened waters in turn pave the way for a sharp slowing of the thermohaline circulation system.

THE PERIOD FROM 2010 TO 2020

*Thermohaline Circulation Collapse*

After roughly sixty years of slow freshening, the thermohaline collapse begins in 2010, disrupting the temperate climate of Europe, which is made possible by the warm flows of the Gulf Stream (the North Atlantic arm of the global thermohaline conveyor). Ocean circulation patterns change, bringing less warm water north and causing an immediate shift in the weather in Northern Europe and eastern North America. The North Atlantic Ocean continues to be affected by fresh water coming from melting glaciers, Greenland's ice sheet, and, perhaps most importantly, increased rainfall and runoff. Decades of high-latitude warming cause increased precipitation and bring additional fresh water to the salty, dense water in the north, which is normally affected mainly by warmer and saltier water from the Gulf Stream. That massive current of warm water no longer reaches far into the North Atlantic. The immediate climatic effect is cooler temperatures in Europe and throughout much of the Northern Hemisphere and a dramatic drop in rainfall in many key agricultural and populated areas. However, the effects of the collapse will be felt in fits and starts, as the traditional weather patterns re-emerge only to be disrupted again—for a full decade.

The dramatic slowing of the thermohaline circulation is anticipated by some ocean researchers, but the United States is not sufficiently prepared for its effects, timing, or intensity. Computer models of the climate and ocean systems, though improved, were unable to produce sufficiently consistent and accurate information for policymakers. As weather patterns shift in the years following the collapse, it is not clear what type of weather future years will bring. While some forecasters believe the cooling and dryness is about to end, others predict a new ice age or a global drought, leaving policy makers and the public highly uncertain about the future climate and what to do, if anything. Is this merely a "blip" of little importance or a fundamental change in the Earth's climate, requiring an urgent massive human response?

COOLER, DRIER, WINDIER CONDITIONS FOR CONTINENTAL
AREAS OF THE NORTHERN HEMISPHERE

## The Weather Report: 2010–2020

- Drought persists for the entire decade in critical agricultural regions and in the areas around major population centers in Europe and eastern North America.
- Average annual temperatures drop by up to five degrees Fahrenheit over Asia and North America and up to six degrees Fahrenheit in Europe.
- Temperatures increase by up to four degrees Fahrenheit in key areas throughout Australia, South America, and southern Africa.
- Winter storms and winds intensify, amplifying the impact of the changes. Western Europe and the North Pacific face enhanced westerly winds.

Each of the years from 2010 to 2020 sees average temperature drops throughout Northern Europe, leading to as much as a six-degrees-Fahrenheit drop in ten years. Average annual rainfall in this region decreases by nearly thirty percent; and winds are up to fifteen percent stronger on average. The climatic conditions are more severe in the continental interior regions of northern Asia and North America.

The effects of the drought are more devastating than the unpleasantness of temperature decreases in the agricultural and populated areas. With the persistent reduction of precipitation in these areas, lakes dry up, river flow decreases, and fresh water supply is squeezed, overwhelming available conservation options and depleting fresh water reserves. The Mega-droughts begin in key regions in Southern China and Northern Europe around 2010 and last throughout the full decade. At the same time, areas that were relatively dry over the past few decades receive persistent years of torrential rainfall, flooding rivers, and regions that traditionally relied on dryland agriculture.

In the North Atlantic region and across northern Asia, cooling is most pronounced in the heart of winter—December, January, and February—although its effects linger through the seasons, the cooling becomes increasingly intense and less predictable. As snow accumulates in mountain regions, the cooling spreads to summertime. In addition to cooling and summertime dryness, wind-pattern velocity strengthens as the atmospheric circulation becomes more zonal.

While weather patterns are disrupted during the onset of the climatic change around the globe, the effects are far more pronounced in Northern Europe for the first five years after the thermohaline circulation collapse. By the second half of this decade, the chill and harsher conditions spread deeper into Southern Europe, North America, and beyond. Northern Europe cools as a pattern of colder weather lengthens the time that sea ice is present over the northern North Atlantic Ocean, creating a further cooling influence and extending the period of wintertime surface air temperatures. Winds pick up as the atmosphere tries to deal with the stronger pole-to-equator temperature gradient. Cold air blowing across the European continent causes especially harsh conditions for agriculture. The combination of wind and dryness causes widespread dust storms and soil loss.

Signs of incremental warming appear in the southern most areas along the Atlantic Ocean, but the dryness doesn't let up. By the end of the decade, Europe's climate is more like Siberia's.

## An Alternative Scenario for the Southern Hemisphere

There is considerable uncertainty about the climate dynamics of the Southern Hemisphere, mainly due to less paleoclimatic data being available than for the Northern Hemisphere. Weather patterns in key regions in the Southern Hemisphere could mimic those of the Northern Hemisphere, becoming colder, drier, and more severe as heat flows from the tropics to the Northern Hemisphere, trying to thermodynamically balance the climatic system. Alternatively, the cooling of the Northern Hemisphere may lead to increased warmth, precipitation, and storms in the south, as the heat normally transported away from equatorial regions by the ocean currents becomes trapped and as greenhouse gas warming continues to accelerate. Either way, it is not implausible that abrupt climate change will bring extreme weather conditions to many of the world's key population and growing regions at the same time—stressing global food, water, and energy supply.

## The Regions: 2010 to 2020

*Europe.* Hit hardest by the climatic change, average annual temperatures drop by six degrees Fahrenheit in under a decade, with more dramatic shifts along the Northwest coast. The climate in northwestern Europe is colder, drier, and windier, making it more like Siberia. Southern Europe experiences less of a change but still suffers from sharp intermittent cooling and rapid temperature shifts. Reduced precipitation causes soil loss to become a problem throughout Europe, contributing to food supply shortages. Europe struggles to stem emigration out of Scandinavian and northern European nations in search of warmth as well as immigration from hard-hit countries in Africa and elsewhere.

*United States.* Colder, windier, and drier weather makes growing seasons shorter and less productive throughout the northeastern United States, and longer and drier in the southwest. Desert areas face increasing windstorms, while agricultural areas suffer from soil loss due to higher wind speeds and reduced soil moisture. The change toward a drier climate is especially pronounced in the southern states. Coastal areas that were at risk during the warming period remain at risk, as rising ocean levels continues along the shores. The United States turns inward, committing its resources to feeding its own population, shoring up its borders, and managing the increasing global tension.

*China.* China, with its high need for food supply given its vast population, is hit hard by a decreased reliability of the monsoon rains. Occasional monsoons during the summer season are welcomed for their precipitation, but have devastating effects as they flood generally denuded land. Longer, colder winters and hotter summers caused by decreased evaporative cooling because of reduced precipitation stress already tight energy and water supplies. Widespread famine causes chaos and internal struggles as a cold and hungry China peers jealously across the Russian and western borders at energy resources.

*Bangladesh.* Persistent typhoons and a higher sea level create storm surges that cause significant coastal erosion, making much of Bangladesh nearly uninhabitable. Further, the rising sea level contaminates fresh water supplies inland, creating a drinking water and humanitarian crisis. Massive emigration occurs, causing tension in China and India, which are struggling to manage the crisis inside their own boundaries.

*East Africa.* Kenya, Tanzania, and Mozambique face slightly warmer weather, but are challenged by persistent drought. Accustomed to dry conditions, these countries were the least influenced by the changing weather conditions, but their food supply is challenged as major grain producing regions suffer.

*Australia.* A major food exporter, Australia struggles to supply food around the globe, as its agriculture is not severely impacted by more subtle changes in its climate. But the large uncertainties about Southern Hemisphere climate change make this benign conclusion suspect.

## Impact on Natural Resources

The changing weather patterns and ocean temperatures affect agriculture, fish and wildlife, water and energy. Crop yields, affected by temperature and water stress as well as length of growing season fall by ten to twenty-five percent and are less predictable as key regions shift from a warming to a cooling trend. As some agricultural pests die due to temperature changes, other species spread more readily due to the dryness and windiness—requiring alternative pesticides or treatment regiments. Commercial fishermen that typically have rights to fish in specific areas will be ill-equipped for the massive migration of their prey.

With only five or six key grain-growing regions in the world (U.S., Australia, Argentina, Russia, China, and India), there is insufficient surplus in global food supplies to offset severe weather conditions in a few regions at the same time—let alone four or five. The world's economic interdependence makes the United States increasingly vulnerable to the economic disruption created by local weather shifts in key agricultural and high population areas around the world. Catastrophic shortages of water and energy supply—both which are stressed around the globe today—cannot be quickly overcome.

## Impact on National Security

Human civilization began with the stabilization and warming of the Earth's climate. A colder, unstable climate meant that humans could neither develop agriculture nor permanent settlements. With the end of the Younger Dryas and the warming and stabilization that followed, humans could learn the rhythms of agriculture and settle in places whose climate was reliably productive. Modern civilization has never experienced weather conditions as persistently disruptive as the ones outlined in this scenario. As a result, the implications for national security outlined in this report are only hypothetical. The actual impacts would vary greatly depending on the nuances of the weather conditions, the adaptability of humanity, and decisions by policy makers.

Violence and disruption stemming from the stresses created by abrupt changes in the climate pose a different type of threat to national security than we are accustomed to today. Military confrontation may be triggered by a desperate need for natural resources such as energy, food and water rather than by conflicts over ideology, religion, or national honor. The shifting motivation for confrontation would alter which countries are most vulnerable and the existing warning signs for security threats.

There is a long-standing academic debate over the extent to which resource constraints and environmental challenges lead to inter-state conflict. While some believe they alone can lead nations to attack one another, others argue that their primary effect is to act as a trigger of conflict among countries that face pre-existing social, economic, and political tension. Regardless, it seems undeniable that severe environmental problems are likely to escalate the degree of global conflict.

Co-founder and President of the Pacific Institute for Studies in Development, Environment, and Security, Peter Gleick outlines the three most fundamental challenges abrupt climate change poses for national security:

1. Food shortages due to decreases in agricultural production.
2. Decreased availability and quality of freshwater due to flooding and droughts.
3. Disrupted access to strategic minerals due to ice and storms.

In the event of abrupt climate change, it's likely that food, water, and energy resource constraints will first be managed through economic, political, and diplomatic means such as treaties and trade embargoes. Over time though, conflicts over land and water use are likely to become more severe—and more violent. As states become increasingly desperate, the pressure for action will grow.

## Decreasing Carrying Capacity

Today, carrying capacity, which is the ability for the Earth and its natural ecosystems, including social, economic, and cultural systems, to support the finite number of people on the planet, is being challenged around the world. According to the International Energy Agency, global demand for oil will grow by 66 percent in the next thirty years, but it's unclear where the supply will come from. Clean water is similarly constrained in many areas around the world. With 815 million people receiving insufficient sustenance worldwide, some would say that as a globe, we're living well above our carrying capacity, meaning there are not sufficient natural resources to sustain our behavior.

Many point to technological innovation and adaptive behavior as a means for managing the global ecosystem. Indeed it has been technological progress that has increased carrying capacity over time. Over centuries we have learned how to produce more food, energy, and access more water. But will the potential of new technologies be sufficient when a crisis like the one outlined in this scenario hits?

Abrupt climate change is likely to stretch carrying capacity well beyond its already precarious limits. And there's a natural tendency or need for carrying capacity to become realigned. As abrupt climate change lowers the world's carrying capacity, aggressive wars are likely to be fought over food, water, and energy. Deaths from war as well as starvation and disease will decrease population size, which, over time, will re-balance with carrying capacity.

When you look at carrying capacity on a regional or state level it is apparent that those nations with a high carrying capacity, such as the United States and Western Europe, are likely to adapt most effectively to abrupt changes in climate, because, relative to their population size, they have more resources to call on. This may give rise to a more severe have, have-not mentality, causing resentment toward those nations with a higher carrying capacity. It may lead to finger-pointing and blame, as the wealthier nations tend to use more energy and emit more greenhouse gasses such as $CO_2$ into the atmosphere. Less important than the scientifically proven relationship between $CO_2$ emissions and climate change is the perception that impacted nations have—and the actions they take.

## The Link between Carrying Capacity and Warfare

Steven LeBlanc, Harvard archaeologist and author of a new book called *Carrying Capacity*, describes the relationship between carrying capacity and warfare. Drawing on abundant archaeological and ethnological data, LeBlanc argues that, historically, humans conducted organized warfare for a variety of reasons, including warfare over resources and the environment. Humans fight when they outstrip the carrying capacity of their natural environment. Every time there is a choice between starving and raiding, humans raid. From hunter/gatherers through agricultural tribes, chiefdoms, and early complex societies, 25 percent of a population's adult males die when war breaks out.

Peace occurs when carrying capacity goes up, as with the invention of agriculture, newly effective bureaucracy, remote trade, and technological breakthroughs. Also a large-scale dieback, such as from plague, can make for peaceful times—Europe after its major plagues, North American natives after European diseases decimated their populations (that's the difference between the Jamestown colony failure and Plymouth Rock success). But such peaceful periods are short-lived because population quickly rises to once again push against carrying capacity, and warfare resumes. Indeed, over the millennia most societies define themselves according to their ability to conduct war, and warrior culture becomes deeply ingrained. The most combative societies are the ones that survive.

However in the last three centuries, LeBlanc points out, advanced states have steadily lowered the body count even though individual wars and genocides have grown larger in scale. Instead of slaughtering all their enemies in the traditional way, for example, states merely kill enough to get a victory and then put the survivors to work in their newly expanded economy. States also use their own bureaucracies, advanced technology, and international rules of behavior to raise carrying capacity and bear a more careful relationship to it.

All of that progressive behavior could collapse if carrying capacities everywhere were suddenly lowered drastically by abrupt climate change. Humanity would revert to its norm of constant battles for diminishing resources, which the battles themselves would further reduce even beyond the climatic effects. Once again warfare would define human life. The two most likely reactions to a sudden drop in carrying capacity due to climate change are defensive and offensive.

The United States and Australia are likely to build defensive fortresses around their countries because they have the resources and reserves to achieve self-sufficiency. With diverse growing climates, wealth, technology, and abundant resources, the United States could likely survive shortened growing cycles and harsh weather conditions without catastrophic losses. Borders will be strengthened around the country to hold back unwanted starving immigrants from the Caribbean islands (an especially severe problem), Mexico, and South America. Energy supply will be shored up through expensive (economically, politically, and morally) alternatives such as nuclear, renewables, hydrogen, and Middle Eastern contracts. Pesky skirmishes over fishing rights, agricultural support, and disaster relief will be commonplace. Tension between the U.S. and Mexico rise as the U.S. reneges on the 1944 treaty that guarantees water flow from the Colorado River. Relief workers will be commissioned to respond to flooding along the southern part of the east coast and much drier conditions inland. Yet, even in this

# CONFLICT SCENARIO DUE TO CLIMATE CHANGE

|  | Europe | Asia | United States |
|---|---|---|---|
| 2010–2020 | 2012: Severe drought and cold push Scandinavian populations southward, push back from EU<br>2015: Conflict within the EU over food and water supply leads to skirmishes and strained diplomatic relations<br>2018: Russia joins EU, providing energy resources<br>2020: Migration from northern countries such as Holland and Germany toward Spain and Italy | 2010: Border skirmishes and conflict in Bangladesh, India, and China, as mass migration occurs toward Burma<br>2012: Regional instability leads Japan to develop force projection capability<br>2015: Strategic agreement between Japan and Russia for Siberia and Sakhalin energy resources<br>2018: China intervenes in Kazakhstan to protect pipelines regularly disrupted by rebels and criminals. | 2010: Disagreements with Canada and Mexico over water increase tension<br>2012: Flood of refugees to southeast U.S. and Mexico from Caribbean islands<br>2015: European migration to United States (mostly wealthy)<br>2016: Conflict with European countries over fishing rights<br>2018: Securing North America, U.S. forms integrated security alliance with Canada and Mexico<br>2020: Department of Defense manages borders and refugees from Caribbean and Europe. |
| 2020–2030 | 2020: Increasing: skirmishes over water and immigration<br>2020: Skirmish between France and Germany over commercial access to Rhine<br>2025: EU nears collapse<br>2027: Increasing migration to Mediterranean countries such as Algeria, Morocco, Egypt, and Israel<br>2030: Nearly 10% of European population moves to a different country | 2020: Persistent conflict in South East Asia; Burma, Laos, Vietnam, India, China<br>2025: Internal conditions in China deteriorate dramatically leading to civil war and border wars.<br>2030: Tension growing between China and Japan over Russian energy | 2020: Oil prices increase as security of supply is threatened by conflicts in Persian Gulf and Caspian<br>2025: Internal struggle in Saudi Arabia brings Chinese and U.S. naval forces to Gulf in direct confrontation |

The chart above outlines some potential military implications of climate change continuous state of emergency, the U.S. will be positioned well compared to others. The intractable problem facing the nation will be calming the mounting military tension around the world.

As famine, disease, and weather-related disasters strike due to the abrupt climate change, many countries' needs will exceed their carrying capacity. This will create a sense of desperation, which is likely to lead to offensive aggression in order to reclaim balance. Imagine Eastern European countries, struggling to feed their populations with a falling supply of food, water, and energy, eyeing Russia, whose population is already in decline, for access to its grain, minerals, and energy supply. Or picture Japan, suffering from flooding along its coastal cities and contamination of its fresh water supply, eying Russia's Sakhalin Island oil and gas reserves as an energy source to power desalination plants and energy-intensive agricultural processes. Envision Pakistan, India, and China—all armed with nuclear weapons—skirmishing at their borders over refugees, access to shared rivers, and arable land. Spanish and Portuguese fishermen might fight over fishing rights—leading to conflicts at sea. And, countries including the United States would be likely to better secure their borders. With over 200 river basins touching multiple nations, we can expect conflict over access to water for drinking, irrigation, and transportation. The Danube touches twelve nations, the Nile runs though nine, and the Amazon runs through seven.

In this scenario, we can expect alliances of convenience. The United States and Canada may become one, simplifying border controls. Or, Canada might keep its hydropower—causing energy problems in the U.S. North and South Korea may align to create one technically savvy and nuclear-armed entity. Europe may act as a unified bloc—curbing immigration problems between European nations—and allowing for protection against aggressors. Russia, with its abundant minerals, oil, and natural gas may join Europe.

In this world of warring states, nuclear arms proliferation is inevitable. As cooling drives up demand, existing hydrocarbon supplies are stretched thin. With a scarcity of energy supply—and a growing need for access—nuclear energy will become a critical source of power, and this will accelerate nuclear proliferation as countries develop enrichment and reprocessing capabilities to ensure their national security. China, India, Pakistan, Japan, South Korea, Great Britain, France, and Germany will all have nuclear weapons capability, as will Israel, Iran, Egypt, and North Korea.

Managing the military and political tension, occasional skirmishes, and threat of war will be a challenge. Countries such as Japan that have a great deal of social cohesion (meaning the government is able to effectively engage its population in changing behavior) are most likely to fair well. Countries whose diversity already produces conflict, such as India, South Africa and Indonesia, will have trouble maintaining order. Adaptability and access to resources will be key. Perhaps the most frustrating challenge abrupt climate change will pose is that we'll never know how far we are into the climate change scenario and how many more years—ten, 100, 1,000—remain before some kind of return to warmer conditions as the thermohaline circulation starts up again. When carrying capacity drops suddenly, civilization is faced with new challenges that today seem unimaginable.

## Could This Really Happen?

Ocean, land, and atmosphere scientists at some of the world's most prestigious organizations have uncovered new evidence over the past decade suggesting that the plausibility of severe and rapid climate change is higher than most of the scientific community and perhaps all of the political community is prepared for. If it occurs, this phenomenon will disrupt current gradual global warming trends, adding to climate complexity and lack of predictability. And paleoclimatic evidence suggests that such an abrupt climate change could begin in the near future.

The Woods Hole Oceanographic Institute reports that seas surrounding the North Atlantic have become less salty in the past forty years, which in turn freshens the deep ocean in the North Atlantic. This trend could pave the way for ocean-conveyor collapse or slowing and abrupt climate change.

With at least eight abrupt climate change events documented in the geological record, it seems that the questions to ask are: When will this happen? What will the impacts be? And how can we best prepare for it? Rather than: Will this really happen?

*Are we prepared for history to repeat itself again?*

There is a debate in newspapers around the globe today on the impact of human activity on climate change. Because economic prosperity is correlated with energy use and greenhouse gas emissions, it is often argued that economic progress leads to climate change. Competing evidence suggests that climate change can occur regardless of human activity, as seen in climate events that happened prior to modern society.

It's important to understand human impacts on the environment—both what's done to accelerate and decelerate (or perhaps even reverse) the tendency toward climate change. Alternative fuels, greenhouse gas emission controls, and conservation efforts are worthwhile endeavors. In addition, we should prepare for the inevitable effects of abrupt climate change—which will likely come regardless of human activity.

Here are some preliminary recommendations to prepare the United States for abrupt climate change:

1. *Improve predictive climate models.* Further research should be conducted so more confidence can be placed in predictions about climate change. There needs to be a deeper understanding of the relationship between ocean patterns and climate change. This research should focus on historical, current, and predictive forces, and aim to further our understanding of abrupt climate change, how it may happen, and how we'll know it's occurring.

2. *Assemble comprehensive predictive models of climate change impacts.* Substantial research should be done on the potential ecological, economic, social, and political impact of abrupt climate change. Sophisticated models and scenarios should be developed to anticipate possible local conditions. A system should be created to identify how climate change may impact the global distribution of social, economic, and political power. These analyses can be used to mitigate potential sources of conflict before they happen.

3. *Create vulnerability metrics.* Metrics should be created to understand a country's vulnerability to the impacts of climate change. Metrics may include climatic impact on existing agricultural, water, and mineral resources; technical capability; social cohesion and adaptability.

4. *Identify no-regrets strategies.* No-regrets strategies should be identified and implemented to ensure reliable access to food supply and water, and to ensure national security.

5. *Rehearse adaptive responses.* Adaptive response teams should be established to address and prepare for inevitable climate-driven events such as massive migration, disease and epidemics, and food and water supply shortages.
6. *Explore local implications.* The first-order effects of climate change are local. While we can anticipate changes in pest prevalence and severity and changes in agricultural productivity, one has to look at very specific locations and conditions to know which pests are of concern, which crops and regions are vulnerable, and how severe impacts will be. Such studies should be undertaken, particularly in strategically important food producing regions.
7. *Explore geo-engineering options that control the climate.* Today, it is easier to warm than to cool the climate, so it might be possible to add various gases, such as hydro-fluorocarbons, to the atmosphere to offset the affects of cooling. Such actions, of course, would be studied carefully, as they have the potential to exacerbate conflicts among nations.

## *Conclusion*

It is quite plausible that within a decade the evidence of an imminent abrupt climate shift may become clear and reliable. It is also possible that our models will better enable us to predict the consequences. In that event the United States will need to take urgent action to prevent and mitigate some of the most significant impacts. Diplomatic action will be needed to minimize the likelihood of conflict in the most impacted areas, especially in the Caribbean and Asia. However, large population movements in this scenario are inevitable. Learning how to manage those populations, border tensions that arise, and the resulting refugees will be critical. New forms of security agreements dealing specifically with energy, food, and water will also be needed. In short, while the U.S. itself will be relatively better off and with more adaptive capacity, it will find itself in a world where Europe will be struggling internally, large numbers of refugees washing up on its shores, and Asia is in serious crisis over food and water. Disruption and conflict will be endemic features of life.

*Arundhati Roy grew up in the Indian state of Kerala. She became famous in 1997 with the publication of her first novel, The God of Small Things. She has devoted most of her energy in the years since to activism, fighting hard against, among other things, the massive Narmada dam. This short and powerful allegory is one of the few pieces of fiction she's written in recent years, but her biting and powerful essays have found a wide audience.*

# The Briefing

from *Field Notes on Democracy: Listening to Grasshoppers*

Arundhati Roy
*2009*

My greetings. I'm sorry I'm not here with you today but perhaps it's just as well. In times such as these, it's best not to reveal ourselves completely, not even to each other.

If you step over the line and into the circle, you may be able to hear better. Mind the chalk on your shoes.

I know many of you have travelled great distances to be here. Have you seen all there is to see? The pillbox batteries, the ovens, the ammunition depots with cavity floors? Did you visit the workers' mass grave? Have you studied the plans carefully? Would you say that it's beautiful, this fort? They say it sits astride the mountains like a defiant lion.

I confess I've never seen it. The guidebook says it wasn't built for beauty. But beauty can arrive uninvited, can it not? It can fall upon things unexpectedly, like sunlight stealing through a chink in the curtains. Ah, but then this is the fort with no chinks in its curtains, the fort that has never been attacked. Does this mean its forbidding walls have thwarted even Beauty and sent it on its way?

Beauty. We could go on about it all day and all night long. What is it? What is it not? Who has the right to decide? Who are the world's real curators, or should we say the real world's curators? What is the real world? Are things we cannot imagine, measure, analyze, represent and reproduce real? Do they exist? Do they live in the recesses of our minds in a fort that has never been attacked? When our imaginations fail, will the world fail too? How will we ever know?

How big is it, this fort that may or may not be beautiful? They say it is the biggest fort ever built in the high mountains.

Gigantic, you say? Gigantic makes things a little difficult for us. Shall we begin by mapping its vulnerabilities? Even though it has never been attacked (or so they say), think of how its creators must have lived and relived the idea of being attacked. They must have waited to be attacked. They must have dreamt of being attacked.

They must have placed themselves in the minds and hearts of their enemies until they could barely tell themselves apart from those they feared so deeply. Until they no longer knew the difference between terror and desire. And then, from that knothole of tormented love, they must have imagined attacks from every conceivable direction with such precision and cunning as to render them almost real. How else could they have built a fortification like this? Fear must have shaped it; dread must be embedded in its very grain. Is that what this fort really is? A fragile testament to trepidation, to apprehension, to an imagination under siege?

It was built—and I quote its chief chronicler—to store everything that ought to be defended at all costs. Unquote. That's saying something. What did they store here comrades? What did they defend?

Weapons. Gold. Civilization itself. Or so the guidebook says.

And now, in Europe's time of peace and plenty, it is being used to showcase the transcendent purpose, or, if you wish, the sublime purposelessness, of civilization's highest aspiration: Art. These days, I'm told, Art is Gold.

I hope you have bought the catalogue. You must. For appearances' sake at least.

As you know, the chances are that there's gold in this Fort. Real gold. Hidden gold. Most of it has been removed, some of it stolen, but a good amount is said to still remain. Everyone's looking for it, knocking on walls, digging up graves. Their urgency must be palpable to you.

They know there's gold in the fort. They also know there's no snow on the mountains. They want the gold to buy some snow.

Those of you who are from here—you must know about the Snow Wars.

Those of you who aren't, listen carefully. It is vital that you understand the texture and fabric of the place you have chosen for your mission.

Since the winters have grown warmer here, there are fewer 'snowmaking' days and as a result there's not enough snow to cover the ski slopes. Most ski slopes can no longer be classified as 'snow-reliable.' At a recent press conference—perhaps you've read the reports—Werner Voltron, president of the Association of Ski Instructors, said, "The future, I think is black. Completely black." (Scattered applause that sounds as though it's coming from the back of the audience. Barely discernible murmurs of *Bravo! Viva! Wah, Wah! Yeah Brother!*) No no no...comrades, comrades...you misunderstand. Mr. Voltron was not referring to the Rise of the Black Nation. By Black he meant ominous, ruinous, hopeless, catastrophic, and bleak. He said that every one degree celsius increase in winter temperatures spells doom for almost one hundred ski resorts. That, as you can imagine, is a lot of jobs and money.

Not everybody is as pessimistic as Mr. Voltron. Take the example of Guenther Holzhausen, CEO of MountainWhite, a new branded snow product, popularly known as Hot Snow (because it can be manufactured at two to three degrees celsius above the normal temperature). Mr. Holzhausen said—and I'll read this out to you—"The changing climate is a great opportunity for the Alps. The extremely high temperatures and rising sea levels brought about by global warming will be bad for seaside tourism. Ten years from now people usually headed for the Mediterranean will be coming to the comparatively cooler Alps for skiing holidays. It is our responsibility; indeed our *duty* to guarantee snow of the highest quality. MountainWhite guarantees dense, evenly spread snow which skiers will find is far superior to natural snow." Unquote.

MountainWhite snow, comrades, like most artificial snows, is made from a protein located in the membrane of a bacterium called Pseudomonas syringae. What sets it apart from other snows is that in order to prevent the spread of disease and other pathogenic hazards, MountainWhite guarantees that the water it uses to generate snow for skiing is of the highest quality, sourced directly from drinking water networks. "You can bottle our ski slopes and drink them!" Guenther Holzhausen is known to have once boasted. (Some restless angry murmuring on the soundtrack.) I understand... But calm your anger. It will only blur your vision and blunt your purpose.

To generate artificial snow, nucleated, treated water is shot out of high-pressure power-intensive snow cannons at high speed. When the snow is ready, it is stacked in mounds called whales. The snow whales are groomed, tilled and fluffed before the snow is evenly spread on slopes that have been shaved of imperfections and natural rock formations. The soil is covered with a thick layer of fertilizer to keep the soil cool and insulate it from the warmth generated by Hot Snow. Most ski resorts use artificial snow now. Almost every resort has a cannon. Every cannon has a brand. Every brand is at war. Every war is an opportunity.

If you want to ski on—or at least *see*—natural snow, you'll have to go further, up to the glaciers that are wrapped in giant sheets of plastic foil to protect them from the summer heat and prevent them from shrinking. I don't know how natural that is though—a glacier wrapped in foil. You might feel as though you're skiing on an old sandwich. Worth a try I suppose. I wouldn't know, I don't ski. The Foil Wars are a form of high-altitude combat—not the kind that some of you are trained for (chuckles). They are separate, though not entirely unconnected to the Snow Wars.

In the Snow Wars, MountainWhite's only serious adversary is Scent n' Sparkle, a new product introduced by Peter Holzhausen, who, if you will pardon me for gossiping, is Guenther Holzhausen's brother.

Real brother. Their wives are sisters. (A murmur) What's that? Yes...real brothers married to real sisters. The families are both from Salzburg.

In addition to all the advantages of MountainWhite, Scent n' Sparkle promises whiter, brighter snow with a fragrance. At a price of course. Scent n' Sparkle comes in three aromas— Vanilla, Pine and Evergreen. It promises to satisfy tourists' nostalgic yearning for old-fashioned holidays. Scent n' Sparkle is a boutique product poised to storm the mass market, or so the pundits say, because it is a product with vision, and an eye to the future. Scented snow anticipates the effects that the global migration of trees and forests will have on the tourism industry. (Murmur) Yes. I did say tree migration.

Did any of you read Macbeth in school? Do you remember what the witches on the heath said to him? *"Macbeth shall never vanquished be, until Great Burnam Wood to high Dunsinane Hill shall come against him?"*

Do you remember what he said to them?

(A voice from the audience somewhere at the back says, *"That will never be. Who can impress the forest, bid the tree unfix his earthbound root?"*)

Ha! Excellent. But Macbeth was dead wrong. Trees *have* unfixed their earthbound roots and are on the move. They're migrating from their devastated homes in the hope of a better life. Like people. Tropical palms are moving up into the Lower Alps. Evergreens are climbing to higher altitudes in search of a colder climate. On the ski slopes, under the damp carpets of Hot Snow, in the warm, fertilizer-coated soil, stowaway seeds of new hothouse plants are germinating. Perhaps soon there'll be fruit trees and vineyards and olive groves in the high mountains.

When the trees migrate, birds and insects, wasps, bees, butterflies, bats and other pollinators will have to move with them. Will they be able to adapt to their new surrounding? Robins have already arrived in Alaska. Alaskan caribou plagued by mosquitoes are moving to higher altitudes where they don't have enough food to eat. Mosquitoes carrying malaria are sweeping through the Lower Alps.

I wonder how this fort that was built to withstand heavy artillery fire will mount a defence against an army of mosquitoes.

The Snow Wars have spread to the plains. MountainWhite now dominates the snow market in Dubai and Saudi Arabia. It is lobbying in India and China, with some success, for dam construction projects dedicated entirely to snow cannons for all-season ski resorts. It has entered the Dutch market for dyke reinforcement and for sea homes built on floating raft foundations, so that when the sea levels rise and the dykes are finally breached and Holland drifts into the ocean, MountainWhite can harness the rising tide and turn it into gold. *Never fear, MountainWhite is here!* Works just as well in the flatlands. Scent n' Sparkle has diversified too. It owns a popular TV channel and controlling shares in a company that makes—as well as defuses—landmines. Perhaps their new batch will be scented—strawberry, cranberry, jojoba—in order to attract animals and birds as well as children. Other than snow and landmines, Scent n' Sparkle also retails mass market, battery-operated, prosthetic limbs in standard sizes for Central Asia and Africa. It is at the forefront of the campaign for Corporate Social Responsibility and is funding a chain of excellently appointed corporate orphanages and NGOs in Afghanistan which some of you are familiar with. Recently it has put in a tender for the dredging and cleaning of lakes and rivers in Austria and Italy that have once again grown toxic from the residue of fertilizer and artificial snowmelt.

Even here, at the top of the world, residue is no longer the past. It is the future. At least some of us have learned over the years to live like rats in the ruins of other people's greed. We have learned to fashion weapons from nothing at all. We know how to use them. These are our combat skills.

Comrades, the stone lion in the mountains has begun to weaken. The Fort that has never been attacked has laid siege to itself. It is time for us to make our move. Time to replace the noisy, undirected spray of machinegun fire with the cold precision of an assassin's bullet. Choose your targets carefully.

When the stone lion's stone bones have been interred in this, our wounded, poisoned earth, when the Fort That Has Never Been Attacked has been reduced to rubble and when the dust from the rubble has settled, who knows, perhaps it will snow again.

That is all I have to say. You may disperse now. Commit your instructions to memory. Go well, comrades, leave no footprints. Until we meet again, godspeed, *khuda hafiz* and keep your powder dry.

(Shuffle of footsteps leaving. Fading away.)

*David Breashears is probably best known for his high-altitude cinematography—a world-class climber, he took the IMAX images for the classic film Everest. But one of his most important projects consists of still images like these. He took old pictures of the roof of the world—many from the 1921 Mallory expedition to Everest—and painstakingly found the same vantage points so he could recreate the shots eight decades later. Side by side, what the images showed was an almost unbelievable loss of ice—the scale of these mountains is so huge that it takes a moment to realize that, in the pictures of the Ronbuk Glacier, 400 vertical feet of ice (that's taller than the Statue of Liberty) has disappeared.*

# Images

David Breashears and E. O. Wheeler

*Top:* "From Kyetrak West." E.O. Wheeler, 1921. Courtesy the Royal Geographical Society
*Above:* Courtesy David Breashears
*Overleaf left:* "West Rongbuk." E.O. Wheeler, 1921. Courtesy the Royal Geographical Society
*Overleaf right:* Courtesy David Breashears

*Vandana Shiva has for decades been a campaigner for local and self-sufficient farming, and against industrialized and genetically engineered agriculture. Here she helps address one of the pressing concerns about climate change: that it will dramatically reduce crop yields. New data from Stanford and the University of Washington have predicted declines as large as forty percent in grain yields as temperature warms, and some have used that to call for aggressive genetic engineering; Shiva suggests that instead we need the resilience offered by peasant agriculture.*

# Climate Change and Agriculture

Dr. Vandana Shiva
*2011*

*Biodiverse Ecological Farming is the Answer, not Genetic Engineering*

Industrial globalized agriculture is heavily implicated in climate change. It contributes to the three major greenhouse gases: carbon dioxide ($CO_2$) from the use of fossil fuels, nitrogen oxide ($N_2O$) from the use of chemical fertilizers and methane ($CH_4$) from factory farming. According to the Intergovernmental Panel on Climate change (IPCC), atmospheric concentration of $CO_2$ has increased from a pre-industrial concentration of about 280 parts per million to 379 parts per million in 2005. The global atmospheric concentration of $CH_4$ has increased from pre-industrial concentration of 715 parts per billion to 1774 parts per billion in 2005. The global atmospheric concentration of $N_2O$, largely due to use of chemical fertilizers in agriculture, increased from about 270 parts per billion to 319 parts per billion in 2005. Industrial agriculture is also more vulnerable to climate change, which is intensifying droughts and floods.

Monocultures lead to more frequent crop failure when rainfall does not come in time, or is too much or too little. Chemically fertilized soils have no capacity to withstand a drought. And cyclones and hurricanes make a food system dependent on long distance transport highly vulnerable to disruption. Genetic engineering is embedded in an industrial model of agriculture based on fossil fuels. It is falsely being offered as a magic bullet for dealing with climate change. Monsanto claims that Genetically Modified Organisms [GMOs] are a cure for both food insecurity and climate change, and has been putting the following advertisement across the world in recent months.

> 9 billion people to feed.
> A changing climate
> Now what?
> Producing more
> Conserving more
> Improving farmers lives
> That's sustainable agriculture
> And that's what Monsanto is all about.

All the claims this advertisement makes are false. GM crops do not produce more. While Monsanto claims its GMO Bt cotton gives 1500 kilograms per acre, the average is 300–400 kilograms per acre. The claim to increased yield is false because yield, like climate resilience is a multi-genetic trait. Introducing toxins into a plant through herbicide resistance or Bt toxin increases the "yield" of toxins, not of food or nutrition. Even the nutrition argument is manipulated. Golden rice genetically engineered to increase Vitamin A produces seventy times less Vitamin A than available alternatives such as coriander leaves and curry leaves. The false claim of higher food production has been dislodged by a recent study titled Failure to Yield by Dr. Doug Gurian Sherman of the Union of Concerned Scientists, who was former biotech specialist for the U.S. Environmental Protection Agency and former adviser on GM to the U.S. Food and Drug Administration. Sherman states, "Let us be clear. There are no commercialized GM crops that inherently increase yield. Similarly there are no GM crops on the market that were engineered to resist drought, reduce fertilizer pollution or save soil. Not one."

There are currently two predominant applications of genetic engineering: one is herbicide resistance, the other is crops with Bt toxin. Herbicides kill plants. Therefore they reduce return of organic matter to the soil. Herbicide-resistant crops, like Round Up Ready Soya and Corn reduce soil carbon, they do not conserve it. This is why Monsanto's attempt to use the climate negotiations to introduce Round Up and Round Up resistant crops as a climate solution is scientifically and ecologically wrong. Monsanto's GMOs, which are either Round Up Ready crops or Bt toxin crops do not conserve resources. They demand more water, they destroy biodiversity, and they increase toxins in farming. Pesticide use has increased 13 times as a result of the use Bt cotton seeds in the region of Vidharbha, India. Monsanto's GMOs do not improve farmers' lives. They have pushed farmers to suicide. Two hundred thousand Indian farmers have committed suicide in the last decade. Eighty-four percent of the suicides in Vidharbha, the region with highest suicides, are linked to debt created by Bt cotton. GMOs are non-renewable, while the open pollinated varieties that farmers have bred are renewable and can be saved year to year. The price of cotton seed was seven rupees per kilogram. Bt cotton seed price jumped to 1,700 rupees per kilogram. This is neither ecological nor economic or social sustainability. It is eco-cide and genocide.

Genetic engineering does not "create" climate resilience. In a recent article titled "GM: Food for Thought" (*Deccan Chronicle*,

August 26, 2009), Dr. M.S. Swaminathan wrote "we can isolate a gene responsible for conferring drought tolerance, introduce that gene into a plant, and make it drought tolerant." Drought tolerance is a polygenetic trait. It is therefore scientifically flawed to talk of "isolating a gene for drought tolerance." Genetic engineering tools are so far only able to transfer single-gene traits. That is why in twenty years only two single-gene traits for herbicide resistance and Bt toxin have been commercialized through genetic engineering. Navdanya's recent report titled, "Biopiracy of Climate Resilient Crops: Gene Giants Are Stealing Farmers' innovation of drought-resistant, flood-resistant and salt-resistant varieties," shows that farmers have bred corps that are resistant to climate extremes. And it is these traits which are the result of millennia of farmers' breeding which are now being patented and pirated by the genetic engineering industry. Using farmers' varieties as "genetic material," the biotechnology industry is playing genetic roulette to gamble on which gene complexes are responsible for which trait. This is not done through genetic engineering; it is done through software programs like Athlete.

As the report states, "Athlete uses vast amounts of available genomic data (mostly public) to rapidly reach a reliable limited list of candidate key genes with high relevance to a target trait of choice. Allegorically, the Athlete platform could be viewed as a 'machine' that is able to choose 50–100 lottery tickets from amongst hundreds of thousands of tickets, with the high likelihood that the winning ticket will be included among them." Breeding is being replaced by gambling; innovation is giving way to biopiracy; and science is being substituted by propaganda. This cannot be the basis of food security in times of climate vulnerability.

While genetic engineering is a false solution, over the past twenty years, we have built Navdanya, India's biodiversity and organic farming movement. We are increasingly realizing there is a convergence between objectives of conservation of biodiversity, reduction of climate change impact, and alleviation of poverty. Biodiverse, local, organic systems produce more food and higher farm incomes, while they also reduce water use and risks of crop failure due to climate change. Biodiversity offers resilience to recover from climate disasters. After the Orissa Super Cyclone of 1998, and the Tsunami of 2004, Navdanya distributed seeds of saline-resistant rice varieties as "Seeds of Hope" to rejuvenate agriculture in lands saline reentered by the sea. We are now creating seed banks of drought-resistant, flood-resistant and saline-resistant seed varieties to respond to climate extremities. Navdanya's work over the past twenty years has shown that we can grow more food and provide higher incomes to farmers without destroying the environment and killing our peasants.

Our study on "Biodiversity based organic farming: A new paradigm for Food Security and Food Safety" has established that small biodiverse organic farms produce more food and provide higher incomes to farmers. Biodiverse organic and local food systems contribute both to mitigation of and adaptation to climate change. Small, biodiverse organic farms, especially in Third World countries, are totally fossil-fuel-free. Energy for farming operations comes from animal energy. Soil fertility is built by feeding soil organisms by recycling organic matter. This reduces greenhouse gas emissions. Biodiverse systems are also more resilient to droughts and floods because they have higher water holding capacity and hence contribute to adaption to climate change. Navdanya's study on climate change and organic farming has indicated that organic farming increases carbon absorption by up to fifty-five percent and water holding capacity by ten percent, thus contributing to both mitigation and adaptation to climate change. Biodiverse organic farms produce more food and higher incomes than industrial monocultures. Mitigating climate change, conserving biodiversity and increasing food security can thus go hand in hand.

One of the recurring challenges to winning action in the fight against global warming is that, at least in the beginning, taking a picture of it was hard. Carbon dioxide, after all, is invisible, and the effects of climate change, while whip-fast on a geological time scale, are slow enough that they're difficult to capture on film. Gary Braasch has been one of the most devoted artists trying to change that reality, and his epic global treks to ravaged coastlines, sinking islands, and dwindling ice sheets give us some of our most powerful images of our new reality.

# Images

Gary Braasch
*2010*

*Above:* Australia dust storm
*Overleaf left:* Drought in Florida
*Overleaf right, top:* Antarctica
*Overleaf right, bottom:* Low water in Lake Powell

*At the time of Hansen's testimony in 1988, and indeed for the decade that followed, no one had thought to wonder about the effects of excess carbon on the chemistry of seawater. The discovery that oceans were quickly acidifying came early in the new millennium, and it came as a shock, since we'd allowed our visceral notion of the sea's vastness to lull us into complacency. Elizabeth Kolbert's remarkable piece for* The New Yorker *helps explain why ocean acidification is now viewed as one of the single most serious effects of excess carbon.*

# The Darkening Sea: What Carbon Emissions Are Doing to the Ocean

Elizabeth Kolbert
2006

Pteropods are tiny marine organisms that belong to the very broad class known as zooplankton. Related to snails, they swim by means of a pair of winglike gelatinous flaps and feed by entrapping even tinier marine creatures in a bubble of mucus. Many pteropod species—there are nearly a hundred in all—produce shells, apparently for protection; some of their predators, meanwhile, have evolved specialized tentacles that they employ much as diners use forks to spear escargot. Pteropods are first male, but as they grow older they become female.

Victoria Fabry, an oceanographer at California State University at San Marcos, is one of the world's leading experts on pteropods. She is slight and soft-spoken, with wavy black hair and blue-green eyes. Fabry fell in love with the ocean as a teen-ager after visiting the Outer Banks, off North Carolina, and took up pteropods when she was in graduate school, in the early nineteen-eighties. At that point, most basic questions about the animals had yet to be answered, and, for her dissertation, Fabry decided to study their shell growth. Her plan was to raise pteropods in tanks, but she ran into trouble immediately. When disturbed, pteropods tend not to produce the mucus bubbles, and slowly starve. Fabry tried using bigger tanks for her pteropods, but the only correlation, she recalled recently, was that the more time she spent improving the tanks, "the quicker they died." After a while, she resigned herself to constantly collecting new specimens. This, in turn, meant going out on just about any research ship that would have her.

Fabry developed a simple, if brutal, protocol that could be completed at sea. She would catch some pteropods, either by trawling with a net or by scuba diving, and place them in one-liter bottles filled with seawater, to which she had added a small amount of radioactive calcium 45. Forty-eight hours later, she would remove the pteropods from the bottles, dunk them in warm ethanol, and pull their bodies out with a pair of tweezers. Back on land, she would measure how much calcium 45 their shells had taken up during their two days of captivity.

In the summer of 1985, Fabry got a berth on a research vessel sailing from Honolulu to Kodiak Island. Late in the trip, near a spot in the Gulf of Alaska known as Station Papa, she came upon a profusion of *Clio pyramidata*, a half-inch-long pteropod with a shell the shape of an unfurled umbrella. In her enthusiasm, Fabry collected too many specimens; instead of putting two or three in a bottle, she had to cram in a dozen. The next day, she noticed that something had gone wrong. "Normally, their shells are transparent," she said. "They look like little gems, little jewels. They're just beautiful. But I could see that, along the edge, they were becoming opaque, chalky."

Like other animals, pteropods take in oxygen and give off carbon dioxide as a waste product. In the open sea, the $CO_2$ they produce has no effect. Seal them in a small container, however, and the $CO_2$ starts to build up, changing the water's chemistry. By overcrowding her *Clio pyramidata*, Fabry had demonstrated that the organisms were highly sensitive to such changes. Instead of growing, their shells were dissolving. It stood to reason that other kinds of pteropods—and, indeed, perhaps any number of shell-building species—were similarly vulnerable. This should have represented a major discovery, and a cause for alarm. But, as is so often the case with inadvertent breakthroughs, it went unremarked upon. No one on the boat, including Fabry, appreciated what the pteropods were telling them, because no one, at that point, could imagine the chemistry of an entire ocean changing.

Since the start of the industrial revolution, humans have burned enough coal, oil, and natural gas to produce some 250 billion metric tons of carbon. The result, as is well known, has been a transformation of the earth's atmosphere. The concentration of $CO_2$ in the air today—380 parts per million—is higher than it has been at any point in the past 650,000 years, and probably much longer. At the current rate of emissions growth, $CO_2$ concentration will top 500 parts per million—roughly double pre-industrial levels—by the middle of this century. It is expected that such an increase will produce an eventual global temperature rise of between three and a half and seven degrees Fahrenheit, and that this, in turn, will prompt a string of disasters, including fiercer hurricanes, more deadly droughts, the disappearance of most remaining glaciers, the melting of the Arctic ice cap, and the inundation of many of the world's major coastal cities. But this is only half the story.

Ocean covers seventy percent of the earth's surface, and everywhere that water and air come into contact there is an exchange. Gases from the atmosphere get absorbed by the ocean and gases dissolved in the water are released into the atmosphere.

When the two are in equilibrium, roughly the same quantities are being dissolved as are getting released. But change the composition of the atmosphere, as we have done, and the exchange becomes lopsided: more $CO_2$ from the air enters the water than comes back out. In the 1990s, researchers from seven countries conducted nearly a hundred cruises, and collected more than 70,000 seawater samples from different depths and locations. The analysis of these samples, which was completed in 2004, showed that nearly half of all the carbon dioxide that humans have emitted since the start of the nineteenth century has been absorbed by the sea.

When $CO_2$ dissolves, it produces carbonic acid, which has the chemical formula $H_2CO_3$. As acids go, $H_2CO_3$ is relatively innocuous—we drink it all the time in Coke and other carbonated beverages—but in sufficient quantities it can change the water's pH. Already, humans have pumped enough carbon into the oceans—some hundred and twenty billion tons—to produce a 0.1 decline in surface pH. Since pH, like the Richter scale, is a logarithmic measure, a 0.1 drop represents a rise in acidity of about thirty percent. The process is generally referred to as "ocean acidification," though it might more accurately be described as a decline in ocean alkalinity. This year alone, the seas will absorb an additional two billion tons of carbon, and next year it is expected that they will absorb another two billion tons. Every day, every American, in effect, adds forty pounds of carbon dioxide to the oceans.

Because of the slow pace of deep-ocean circulation and the long life of carbon dioxide in the atmosphere, it is impossible to reverse the acidification that has already taken place. Nor is it possible to prevent still more from occurring. Even if there were some way to halt the emission of $CO_2$ tomorrow, the oceans would continue to take up carbon until they reached a new equilibrium with the air. As Britain's Royal Society noted in a recent report, it will take "Tens of thousands of years for ocean chemistry to return to a condition similar to that occurring at pre-industrial times."

Humans have, in this way, set in motion change on a geologic scale. The question that remains is how marine life will respond. Though oceanographers are just beginning to address the question, their discoveries, at this early stage, are disturbing. A few years ago, Fabry finally pulled her cloudy shells out of storage to examine them with a scanning electron microscope. She found that their surfaces were riddled with pits. In some cases, the pits had grown into gashes, and the upper layer had started to pull away, exposing the layer underneath.

The term "ocean acidification" was coined in 2003 by two climate scientists, Ken Caldeira and Michael Wickett, who were working at the Lawrence Livermore National Laboratory, in Northern California. Caldeira has since moved to the Carnegie Institution, on the campus of Stanford University, and during the summer I went to visit him at his office, which is housed in a "green" building that looks like a barn that has been taken apart and reassembled at odd angles. The building has no air-conditioning; temperature control is provided by a shower of mist that rains down into a tiled chamber in the lobby. At the time of my visit, California was in the midst of a record-breaking heat wave; the system worked well enough that Caldeira's office, if not exactly cool, was at least moderately comfortable.

Caldeira is a trim man with wiry brown hair and a boyish sort of smile. In the 1980s, he worked as a software developer on Wall Street, and one of his clients was the New York Stock

Exchange, for whom he designed computer programs to help detect insider trading. The programs functioned as they were supposed to, but after a while Caldeira came to the conclusion that the N.Y.S.E. wasn't actually interested in catching insider traders, and he decided to switch professions. He went back to school, at N.Y.U., and ended up becoming a climate modeler.

Unlike most modelers, who focus on one particular aspect of the climate system, Caldeira is, at any given moment, working on four or five disparate projects. He particularly likes computations of a provocative or surprising nature; for example, not long ago he calculated that cutting down all the world's forests and replacing them with grasslands would have a slight cooling effect. (Grasslands, which are lighter in color than forests, absorb less sunlight.) Other recent calculations that Caldeira has made show that to keep pace with the present rate of temperature change plants and animals would have to migrate poleward by thirty feet a day, and that a molecule of $CO_2$ generated by burning fossil fuels will, in the course of its lifetime in the atmosphere, trap a hundred thousand times more heat than was released in producing it.

Caldeira began to model the effects of carbon dioxide on the oceans in 1999, when he did some work for the Department of Energy. The department wanted to know what the environmental consequences would be of capturing $CO_2$ from smokestacks and injecting it deep into the sea. Caldeira set about calculating how the ocean's pH would change as a result of deep-sea injection, and then compared that result with the current practice of pouring carbon dioxide into the atmosphere and allowing it to be taken up by surface waters. In 2003, he submitted his work to *Nature*. The journal's editors advised him to drop the discussion of deep-ocean injection, he recalled, because the calculations concerning the effects of ordinary atmospheric release were so startling. Caldeira published the first part of his paper under the subheading "The coming centuries may see more ocean acidification than the past 300 million years."

Caldeira told me that he had chosen the term "ocean acidification" quite deliberately, for its shock value. Seawater is naturally alkaline, with a pH ranging from 7.8 to 8.5—a pH of 7 is neutral—which means that, for now, at least, the oceans are still a long way from actually turning acidic. Meanwhile, from the perspective of marine life, the drop in pH matters less than the string of chemical reactions that follow.

The main building block of shells is calcium carbonate—$CaCO_3$. (The White Cliffs of Dover are a huge $CaCO_3$ deposit, the remains of countless tiny sea creatures that piled up during the Cretaceous—or "chalky"—period.) Calcium carbonate produced by marine organisms comes in two principal forms, aragonite and calcite, which have slightly different crystal structures. How, exactly, different organisms form calcium carbonate remains something of a mystery. Ordinarily in seawater, $CaCO_3$ does not precipitate out as a solid. To build their shells, calcifying organisms must, in effect, assemble it. Adding carbonic acid to the water complicates their efforts, because it reduces the number of carbonate ions in circulation. In scientific terms, this is referred to as "lowering the water's saturation state with respect to calcium carbonate." Practically, it means shrinking the supply of material available for shell formation. (Imagine trying to build a house when someone keeps stealing your bricks.) Once the carbonate concentration gets pushed low enough, even existing shells, like those of Fabry's pteropods, begin to dissolve.

To illustrate, in mathematical terms, what the seas of the future will look like, Caldeira pulled out a set of graphs. Plotted on one axis was aragonite saturation levels; on the other, latitude. (Ocean latitude is significant because saturation levels tend naturally to decline toward the poles.) Different colors of lines represented different emissions scenarios. Some scenarios project that the world's economy will continue to grow rapidly and that this growth will be fueled mostly by oil and coal. Others assume that the economy will grow more slowly, and still others that the energy mix will shift away from fossil fuels. Caldeira considered four much-studied scenarios, ranging from one of the most optimistic, known by the shorthand B1, to one of the most pessimistic, A2. The original point of the graphs was to show that each scenario would produce a different ocean. But they turned out to be more similar than Caldeira had expected.

Under all four scenarios, by the end of this century the waters around Antarctica will become undersaturated with respect to aragonite—the form of calcium carbonate produced by pteropods and corals. (When water becomes undersaturated, it is corrosive to shells.) Meanwhile, surface pH will drop by another 0.2, bringing acidity to roughly double what it was in pre-industrial times. To look still further out into the future, Caldeira modeled what would happen if humans burned through all the world's remaining fossil-fuel resources, a process that would release some 18,000 gigatons of carbon dioxide. He found that by 2300 the oceans would become undersaturated from the poles to the equator. Then he modeled what would happen if we pushed still further and burned through unconventional fuels, like low-grade shales. In that case, we would drive the pH down so low that the seas would come very close to being acidic.

"I used to think of B1 as a good scenario, and I used to think of A2 as a terrible scenario," Caldeira told me. "Now I look at them as different flavors of bad scenarios."

He went on, "I think there's a whole category of organisms that have been around for hundreds of millions of years which are at risk of extinction—namely, things that build calcium-carbonate shells or skeletons. To a first approximation, if we cut our emissions in half it will take us twice as long to create the damage. But we'll get to more or less the same place. We really need an order-of-magnitude reduction in order to avoid it."

Caldeira said that he had recently gone to Washington to brief some members of Congress. "I was asked, 'What is the appropriate stabilization target for atmospheric $CO_2$?' " he recalled. "And I said, 'Well, I think it's inappropriate to think in terms of stabilization targets. I think we should think in terms of emissions targets.' And they said, 'OK, what's the appropriate emissions target?' And I said, 'Zero.'

"If you're talking about mugging little old ladies, you don't say, 'What's our target for the rate of mugging little old ladies?' You say, 'Mugging little old ladies is bad, and we're going to try to eliminate it.' You recognize you might not be a hundred percent successful, but your goal is to eliminate the mugging of little old ladies. And I think we need to eventually come around to looking at carbon-dioxide emissions the same way."

Coral reefs grow in a great swath that stretches like a belt around the belly of the earth, from thirty degrees north to thirty degrees south latitude. The world's largest reef is the Great Barrier, off the coast of northeastern Australia, and the second largest is off the coast of Belize. There are extensive coral reefs in the tropical Pacific, in the Indian Ocean, and in the Red Sea, and many smaller ones in the Caribbean. These reefs, home to an estimated twenty-five percent of all marine fish species, represent some of the most diverse ecosystems on the planet.

Much of what is known about coral reefs and ocean acidification was originally discovered, improbably enough, in Arizona, in the self-enclosed, supposedly self-sufficient world known as Biosphere 2. A three-acre glassed-in structure shaped like a ziggurat, Biosphere 2 was built in the late 1980s by a private group—a majority of the funding came from the billionaire Edward Bass—and was intended to demonstrate how life on earth (Biosphere 1) could be re-created on, say, Mars. The building contained an artificial "ocean," a "rain forest," a "desert," and an "agricultural zone." The first group of Biosphereans—four men and four women—managed to remain, sealed inside, for two years. They produced all their own food and, for a long stretch, breathed only recycled air, but the project was widely considered a failure. The Biosphereans spent much of the time hungry, and, even more ominously, they lost control of their artificial atmosphere. In the various "ecosystems," decomposition, which takes up oxygen and gives off $CO_2$, was supposed to be balanced by photosynthesis, which does the reverse. But, for reasons mainly having to do with the richness of the soil that had been used in the "agricultural zone," decomposition won out. Oxygen levels inside the building kept falling, and the Biosphereans developed what amounted to altitude sickness. Carbon-dioxide levels soared, at one point reaching three thousand parts per million, or roughly eight times the levels outside.

When Biosphere 2 officially collapsed, in 1995, Columbia University took over the management of the building. The University's plan was to transform it into a teaching and research facility, and it fell to a scientist named Chris Langdon to figure out something pedagogically useful to do with the "ocean," a tank the size of an Olympic swimming pool. Langdon's specialty was measuring photosynthesis, and he had recently finished a project, financed by the Navy, that involved trying to figure out whether blooms of bioluminescent algae could be used to track enemy submarines. (The answer was no.) Langdon was looking for a new project, but he wasn't sure what the "ocean" was good for. He began by testing various properties of the water. As would be expected in such a high-$CO_2$ environment; he found that the pH was low.

"The very first thing I did was try to establish normal chemistry," he recalled recently. "So I added chemicals—essentially baking soda and baking powder—to the water to bring the pH back up." Within a week, the alkalinity had dropped again, and he had to add more chemicals. The same thing happened. "Every single time I did it, it went back down, and the rate at which it went down was proportional to the concentration. So, if I added more, it went down faster. So I started thinking, What's going on here? And then it dawned on me."

Langdon left Columbia in 2004 and now works at the Rosenstiel School of Marine and Atmospheric Science, at the University of Miami. He is fifty-two with a high forehead, deep-set eyes, and a square chin. When I went to visit him, not long ago, he took me to see his coral samples, which were growing in a sort of aquatic nursery across the street from his office. On the way, we had to pass through a room filled with tanks of purple sea slugs, which were being raised for medical research. In the front row, the youngest sea slugs, about half an inch long, were floating gracefully, as if suspended in gelatin. Toward the back were slugs that had been fed for several months on a lavish experimental diet. These were the size of my forearm and seemed barely able to lift their knobby, purplish heads.

Langdon's corals were attached to tiles arranged at the bottom of long, sinklike tanks. These were hundreds of them, grouped by species: *Acropora cervicornis*, a type of staghorn coral that grows in a classic antler shape; *Montastrea cavernosa*, a coral that looks like a seafaring cactus; and *Porites divaricata*, a branching coral made up of lumpy, putty-colored protuberances. Water was streaming into the tanks, but when Langdon put his hand in front of the faucet to stop the flow, I could see that every lobe of *Porites divaricata* was covered with tiny pink arms and that every arm ended in soft, fingerlike tentacles. The arms were waving in what looked to be a frenzy either of joy or of supplication.

Langdon explained that the arms belonged to separate coral polyps, and that a reef consisted of thousands upon thousands of polyps spread, like a coating of plaster, over a dead calcareous skeleton. Each coral polyp is a distinct individual, with its own tentacles and its own digestive system, and houses its own collection of symbiotic algae, known as zooxanthellae, which provide it with most of its nutrition. At the same time, each polyp is joined to its neighbors through a thin layer of connecting tissue, and all are attached to the colony's collective skeleton. Individual polyps constantly add to the group skeleton by combining calcium and carbonate icons in a medium known as the extracytoplasmic calcifying fluid. Meanwhile, other organisms, like parrot fish and sponges, are constantly eating away at the reef in search of food or protection. If a reef were ever to stop calcifying, it would start to shrink and eventually would disappear.

"It's just like a tree with bugs," Langdon explained. "It needs to grow pretty quickly just to stay even."

As Langdon struggled, unsuccessfully, to control the pH in the Biosphere "ocean," he started to wonder whether the corals in the tank might be to blame. The Biosphereans had raised twenty different species of coral, and while many of the other creatures, including nearly all the vertebrates selected for the project, had died out, the corals had survived. Langdon wondered whether the chemicals he was adding to raise the pH were, by increasing the saturation state, stimulating their growth. At the time, it seemed an unlikely hypothesis, because the prevailing view among marine biologists was that corals weren't sensitive to changes in saturation. (In many textbooks, the formula for coral calcification is still given incorrectly, which helps explain the prevalence of this view.) Just about everyone, including Langdon's own post-doc, a young woman named Francesca Marubini, thought that his theory was wrong. "It was a total pain in the ass," Langdon recalled.

To test his hypothesis, Langdon employed a straightforward but time-consuming procedure. Conditions in the "ocean" would be systematically varied, and the growth of the coral monitored. The experiment took more than three years to complete, produced more than a thousand measurements, and, in the end, confirmed Langdon's hypothesis. It revealed a more or less linear relationship between how fast the coral grew and how highly saturated the water was. By proving that increased saturation spurs coral growth, Langdon also, of course, demonstrated the reverse: when saturation drops, coral growth slows. In the artificial world of Biosphere 2, the implications of this discovery were interesting; in the real world they were rather more grim. Any drop in the ocean's saturation levels, it seemed, would make coral more vulnerable.

Langdon and Marubini published their findings in the journal *Global Biogeochemical Cycles* in the summer of 2000. Still, many marine biologists remained skeptical, in no small part, it seems, because of the study's association with the discredited Biosphere project. In 2001, Langdon sold his house in New York and moved to Arizona. He spent another two years redoing the experiments, with even stricter controls. The results were essentially identical. In the meantime, other researchers launched similar experiments on different coral species. Their findings were also the same, which, as Langdon put it to me, "is the best way to make believers out of people."

Coral reefs are under threat for a host of reasons: bottom trawling, dynamite fishing, coastal erosion, agricultural runoff, and, nowadays, global warming. When water temperatures rise too high, corals lose—or perhaps expel, no one is quite sure—the algae that nourish them. (The process is called "bleaching," because without their zooxanthellae corals appear white.) For a particular reef, any one of these threats could potentially be fatal. Ocean acidification poses a different kind of threat, one that could preclude the very possibility of a reef.

Saturation levels are determined using a complicated formula that involves multiplying the calcium and carbonate ion concentrations, and then dividing the result by a figure called the stoichiometric solubility product. Prior to the industrial revolution, the world's major reefs were all growing in water whose aragonite saturation level stood between 4 and 5. Today, there is not a single remaining region in the oceans where the saturation level is above 4.5, and there are only a handful of spots—off the northeastern coast of Australia, in the Philippine Sea, and near the Maldives—where it is above 4. Since the takeup of $CO_2$ by the oceans is a highly predictable physical process, it is possible to map the saturation levels of the future with great precision. Assuming that current emissions trends continue, by 2060 there will be no regions left with a level above 3.5. By 2100, none will remain above 3.

As saturation levels decline, the rate at which reefs add aragonite through calcification and the rate at which they lose it through bioerosion will start to approach each other. At a certain point, the two will cross, and reefs will begin to disappear. Precisely where that point lies is difficult to say, because erosion may well accelerate as ocean pH declines. Langdon estimates that the crossing point will be reached when atmospheric $CO_2$ levels exceed six hundred and fifty parts per million, which, under a "business as usual" emissions scenario, will occur sometime around 2075.

"I think that this is just an absolute limit, something they can't cope with," he told me. Other researchers put the limit somewhat higher, and others somewhat lower.

Meanwhile, as global temperatures climb, bleaching events are likely to become more common. A major worldwide bleaching event occurred in 1998, and many Caribbean reefs suffered from bleaching again during the summer of 2005. Current conditions in the equatorial Pacific suggest that 2007 is apt to be another bleaching year. Taken together, acidification and rising ocean temperatures represent a kind of double bind for reefs: regions that remain hospitable in terms of temperature are becoming increasingly inhospitable in terms of saturation, and vice versa.

"While one, bleaching, is an acute stress that's killing them off, the other, acidification, is a chronic stress that's preventing them from recovering," Joanie Kleypas, a reef scientist at the National Center for Atmospheric Research, in Boulder, Colorado, told me. Kleypas said she thought that some corals would be able to migrate to higher latitudes as the oceans warm, but that, because of the lower saturation levels, as well as the difference in light regimes, the size of these migrants would be severely limited. "There's a point where you're going to have coral but no reefs," she said.

The tropical oceans are, as a rule, nutrient-poor; they are sometimes called liquid deserts. Reefs are so dense with life that they are often compared to rain forests. This rain-forest-in-the-desert effect is believed to be a function of a highly efficient recycling system, through which nutrients are, in effect, passed from one reef-dwelling organism to another. It is estimated that at least a million, and perhaps as many as nine million, distinct species live on or near reefs.

"Being conservative, let's say it's a million species that live in and around coral," Ove Hoegh-Guldberg, an expert on coral reefs at the University of Queensland, in Australia, told me. "Some of these species that hang around coral reefs can sometimes be found living without coral. But most species are completely dependent on coral—they literally live in, eat, and breed around coral. And, when we see coral get destroyed during bleaching events, those species disappear. The key question is how vulnerable all these various species are. That's a very important question, but at the moment you'd have to say that a million different species are under threat."

He went on, "This is a matter of the utmost importance. I can't really stress it in words strong enough. It's a do-or-die situation."

Around the same time that Langdon was performing his coral experiments at the Biosphere, a German marine biologist named Ulf Riebesell decided to look into the behavior of a class of phytoplankton known as coccolithophores. Coccolithophores build plates of calcite—coccoliths—that they arrange around themselves, like armor, in structures known as coccospheres. (Viewed under an electron microscope, they look like balls that have been covered with buttons.) Coccolithophores are very tiny—only a few microns in diameter—and also very common. One of the species that Riebesell studied, *Emiliani huxleyi*, produces blooms that can cover forty thousand square miles, turning vast sections of the ocean an eerie milky blue.

In his experiments, Riebesell bubbled $CO_2$ into tanks of coccolithophores to mimic the effects of rising atmospheric concentrations. Both of the species he was studying—*Emiliani huxleyi* and *Gephyrocapsa oceanica*—showed a clear response to the variations. As $CO_2$ levels rose, not only did the organisms' rate of calcification slow; they also started to produce deformed coccoliths and ill-shaped coccospheres.

"To me, it says that we will have massive changes," Riebesell, who works at the Leibniz Institute of Marine Sciences, in Kiel, told me. "If a whole group of calcifiers drops out, are there other organisms taking their place? What is the rate of evolution to fill those spaces? That's awfully difficult to address in experimental work. These organisms have never, ever seen this in their entire evolutionary history. And if they've never seen it they probably will find it difficult to deal with."

Calcifying organisms come in a fantastic array of shapes, sizes, and taxonomic groups. Echinoderms like starfish are calcifiers. So are mollusks like clams and oysters, and crustaceans like barnacles, and many species of bryozoans, or sea mats, and tiny protists known as foraminifera—the list goes on and on. Without experimental data, it's impossible to know which species will prove to be particularly vulnerable to declining pH and which will not. In the natural world, the pH of the water changes by season, and even time of day, and many species may be able to adapt to new conditions, at least within certain bounds. Obviously, though, it's impractical to run experiments on tens of thousands of different species. (Only a few dozen have been tested so far.) Meanwhile, as the example of coral reefs makes clear, what's more important than how acidification will affect any particular organism is how it will affect entire marine ecosystems—a question that can't be answered by even the most ambitious experimental protocol. The recent report on acidification by Britain's Royal Society noted that it was "not possible to predict" how whole communities would respond, but went on to observe that "without significant action to reduce $CO_2$ emissions" there may be "no place in the future oceans for many of the species and ecosystems we know today."

Carol Turley is a senior scientist at Plymouth Marine Laboratory, in Plymouth, England, and one of the authors of the Royal Society report. She observed that pH is a critical variable not just in calcification but in other vital marine processes, like the cycling of nutrients.

"It looks like we'll be changing lots of levels in the food chain," Turley told me. "So we may be affecting the primary producers. We may be affecting larvae of zooplankton and so on. What I think might happen, and it's pure speculation, is that you may get a shortening of the food chain so that only one or two species comes out on top—for instance, we may see massive blooms of jellyfish and things like that, and that's very short food chain."

Thomas Lovejoy, who coined the term "biological diversity" in 1980, compared the effects of ocean acidification to "running the course of evolution in reverse."z

"For an organism that lives on land, the two most important factors are temperature and moisture," Lovejoy, who is now the president of the Heinz Center for Science, Economics, and the Environment, in Washington, D.C, told me. "And for an organism that lives in the water the two most important factors are temperature and acidity. So this is just a profound, profound change. It is going to send all kinds of ripples through marine ecosystems, because of the importance of calcium carbonate for so many organisms in the oceans, including those at the base of the food chain. If you back off and look at it, it's as if you or I went to our annual physical and the body chemistry came back and the doctor looked really, really worried. It's a systemic charge. You could have food chains collapse, and fisheries ultimately with them, because most of the fish we get from the ocean are at the end of long food chains. You probably will see shifts in favor of invertebrates, or the reign of jellyfish."

Riebesell put it this way: "The risk is that at the end we will have the rise of slime."

Paleo-oceanographers study the oceans of the geologic past. For the most part, they rely on sediments pulled up from the bottom of the sea, which contain what might be thought of as a vast library written in code. By analyzing the oxygen isotopes of ancient shells, paleo-oceanographers can, for example, infer the temperature of the oceans going back at least a hundred million years, and also determine how much—or how little—of the planet was covered by ice. By analyzing mineral grains and deposits of "microfossils," they can map archaic currents and wind patterns, and by examining the remains of foraminifera they can re-create the history of ocean pH.

In September, two dozen paleo-oceanographers met with a roughly equal number of marine biologists at a conference hosted by Columbia University's Lamont-Doherty Earth Observatory. The point of the conference, which was titled "Ocean Acid-ification—Modern Observations and Past Experiences," was to use the methods of paleo-oceanography to look into the future.

(The ocean-acidification community is still a relatively small one, and at the conference I ran into half the people I had spoken to about the subject, including Victoria Fabry, Ken Caldeira, and Chris Langdon.) Most of the meeting's first day was devoted to a discussion of an ecological crisis known as the Paleocene-Eocene Thermal Maximum, or P.E.T.M.

The P.E.T.M. took place fifty-five million years ago, at the border marking the end of the Paleocene epoch and the beginning of the Eocene, when there was a sudden, enormous release of carbon into the atmosphere. After the release, temperatures around the world soared; the Arctic, for instance, warmed by ten degrees Fahrenheit, and Antarctica became temperate. Presumably because of this, vertebrate evolution veered off in a new direction. Many of the so-called archaic mammals became extinct, and were replaced by entirely new orders: the ancestors of today's deer, horses, and primates all appeared right around the time of the P.E.T.M. The members of these new orders were curiously undersized—the earliest horse was no bigger than a poodle—a function, it is believed, of hot, dry conditions that favored smallness.

In the oceans, temperatures rose dramatically and, because of all the carbon, the water became increasingly acidic. Marine sediments show that many calcifying organisms vanished—more than fifty species of foraminifera, for example, died out—while others that were once rare became dominant. On the sea floor, the usual buildup of empty shells from dead calcifiers ceased. In ocean cores, the P.E.T.M. shows up vividly as a band of reddish clay sandwiched between thick layers of calcium carbonate.

No one is sure exactly where the carbon of the P.E.T.M. came from or what triggered its release. (Deposits of natural gas known as methane hydrates, which sit, frozen, underneath the ocean floor, are one possible source.) In all, the release amounted to about two trillion metric tons, or eight times as much carbon as humans have added to the atmosphere since industrialization began. This is obviously a significant difference in scale, but the consensus at the conference was that if there was any disparity between then and now it was that the impact of the P.E.T.M. was not drastic enough.

The seas have a built-in buffering capacity: if the water's pH starts to drop, shells and shell fragments that have been deposited on the ocean floor begin to dissolve, pushing the pH back up again. This buffering mechanism is highly effective, provided that acidification takes place on the same timescale as deep-ocean circulation. (One complete exchange of surface and bottom water takes thousand of years.) Paleo-oceanographers estimate that the release of carbon during the P.E.T.M. took between one and ten thousand years—the record is not detailed enough to be more exact—and thus occurred too rapidly to be completely buffered. Currently, $CO_2$ is being released into the air at least three times and perhaps as much as thirty times as quickly as during the P.E.T.M. This is so fast that buffering by ocean sediments is not even a factor.

"In our case, the surface layer is bearing all the burden," James Zachos, a paleo-oceanographer at the University of California at Santa Cruz, told me. "If anything you can look at the P.E.T.M. as a best-case scenario." Ken Caldeira said that he thought a better analogy for the future would be the so-called K-T, or Cretaceous-Tertiary, boundary event, which occurred sixty-five million years ago, when an asteroid six miles wide hit the earth. In addition to dust storms, fires, and tidal waves, the impact is believed to have generated huge quantities of sulfuric acid.

"The K-T boundary event was more extreme but shorter-lived than what we could do in the coming centuries," Caldeira said. "But by the time we've burned conventional fossil-fuel resources what we've done will be comparable in extremeness, except that it will last millennia instead of years." More than a third of all marine genera disappeared at the K-T boundary. Half of all coral species became extinct, and it took the other half more than two million years to recover.

Ultimately, the seas will absorb most of the $CO_2$ that humans emit. (Over the very long term, the figure will approach ninety percent.) From a certain vantage point, this is a lucky break. Were the oceans not providing a vast carbon sink, almost all of the $CO_2$ that humans have emitted would still be in the air. Atmospheric concentrations would now be nearing five hundred parts per million, and the disasters predicted for the end of the century would already be upon us. That there is still a chance to do something to avert the worst consequences of global warming is thanks largely to the oceans.

But this sort of accounting may be misleading. As the process of ocean acidification demonstrates, life on land and life in the seas can affect each other in unexpected ways. Actions that might appear utterly unrelated—say, driving a car down the New Jersey Turnpike and secreting a shell in the South Pacific—turn out to be connected. To alter the chemistry of the seas is to take a very large risk, and not just with the oceans.

*Probably no newspaper, at least in the English-language world, has covered climate change as thoroughly and accurately as London's Guardian, thanks in large part to John Vidal. A veteran reporter and the paper's environment editor since 1995, he has paid attention not only to the science but also to the human effects of climate change, traveling the planet to visit the people already deeply affected by rising temperatures and shifting seasons. His reports are straightforward but heartrending, and it's always hard to read them without reflecting on one of the hideous realities of global warming: it strikes first and hardest at those who have done the least to cause it, and who have the fewest defenses against it.*

# Nepal's Farmers on the Front Line of Global Climate Change
Himalayan communities face catastrophic floods as weather patterns alter

John Vidal
2006

Schoolteacher Sherbahadur Tamang walks through the southern Nepalese village of Khetbari and describes what happened on September 9: "During the night there was light rain but when we woke, its intensity increased. In an hour or so, the rain became so heavy that we could not see more than a foot or two in front of us. It was like a wall of water and it sounded like 10,000 lorries. It went on like that until midday. Then all the land started moving like a river."

When it stopped raining Mr. Tamang and the village barely recognized their valley in the Chitwan hills. In just six hours the Jugedi River, which normally flows for only a few months of the year and is at most about fifty meters wide in Khetbari, had scoured a 300 metre-wide path down the valley, leaving a three meter-deep rockscape of giant boulders, trees and rubble in its path. Hundreds of fields and terraces had been swept away. The irrigation systems built by generations of farmers had gone and houses were demolished or were now uninhabitable. Mr. Tamang's house was left on a newly formed island.

Khetbari expects a small flood every decade or so, but what shocked the village was that the two largest have taken place in the last three years. According to Mr. Tamang, a pattern is emerging. "The floods are coming more severely more frequently. Not only is the rainfall far heavier these days than anyone has ever experienced, it is also coming at different times of the year."

Nepal is on the front line of climate change and variations on Khetbari's experience are now being recorded in communities from the freezing Himalayas of the north to the hot lowland plains of the south. For some people the changes are catastrophic.

"The rains are increasingly unpredictable. We always used to have a little rain each month, but now when there is rain it's very different. It's more concentrated and intense. It means that crop yields are going down," says Tekmadur Majsi, whose lands have been progressively washed away by the Tirshuli River. He now lives with two hundred other environmental change refugees in tents in a small grove of trees by a highway. In the south, villagers are full of minute observations of a changing climate. One notes that wild pigs in the forest now have their young earlier, another that certain types of rice and cucumber will no longer grow where they used to, a third that the days are hotter and that some trees now flower twice a year.

Anecdotal observations are backed by scientists who are recording in Nepal some of the fastest long-term increases in temperatures and rainfall anywhere in the world. At least forty-four of Nepal's and neighbouring Bhutan's Himalayan lakes, which collect glacier meltwater, are said by the UN to be growing so rapidly they could burst their banks within a decade. Any climate change in Nepal is reflected throughout the region. Nearly four hundred million people in northern India and Bangladesh also depend on rainfall and rivers that rise there.

"Unless the country learns to adapt then people will suffer greatly," says Gehendra Gurung, a team leader with Practical

Action in Nepal, which is trying to help people prepare for change. In projects around the country the organization is working with vulnerable villages, helping them build dykes and set up early warning systems. It is also teaching people to grow new crops, introducing drip irrigation and water storage schemes, trying to minimize deforestation which can lead to landslides and introducing renewable energy.

Some people are learning fast and are benefiting. Davandrod Kardigardi, a farmer in the Chitwan village of Bharlang, was taught to grow fruit and, against his father's advice, planted many banana trees. It has paid off handsomely. As other farmers have struggled he has increased his income.

But Nepal as a country needs help adapting to climate change, says Mr. Gurung. Its emissions of damaging greenhouse gases are negligible, yet it finds itself on the front line of change.

"Western countries can control their emissions but to mitigate the effects will take a long time. Until then they can help countries like Nepal to adapt. But it means everyone must question the way they live," he says.

# REFERENCES

### On the Influence of Carbonic Acid in the Air upon the Temperature of the Ground
*Svante Arrhenius*
p. 19

1. For details *cf.* Neumayr, *Erdgeschichte*, Bd.2, Leipzig, 1887; and Geikie, "The Great Ice-Age," 3rd ed. London, 1894; Nathorst, *Jordens historia*, p. 989, Stockholm, 1894.
2. Neumayr, *Erdgeschichte*, p. 648; Nathorst, *l.c.* p. 992.
3. Högbom, *Svensk kemisk Tidskrift*, Bd. vi. p. 169 (1894). *Phil. Mag. S.* Vol. 41. No.251. April 1896. U
4. Luigi De Marchi: *Le cause dell' era glaciale*, premiato dal R. Istituto Lombardo, Pavia, 1895.
5. De Marchi, *l. c.* p. 166.

### The Artificial Production of Carbon Dioxide and Its Influence on Temperature
*G. S. Callendar*
p. 33

Angstrom, A., 1918: *Smithson. Misc. Coll.,* 65, No. 3.
Arrhenius, Svante, 1903: *Kosmische Physik,* 2.
Brown, H., and Escombe, F., 1905: *Proc. Roy. Soc,* B, 76.
Brunt, D., 1932: *Quart. J.R. Met. Soc,* 58.
Carpenter, T. M., 1937: *J. Amer. Chem. Soc.,* 59.
Dines, W. H., 1927: *Mem. R. Met. Soc.,* 2, No. 11.
Fox, C. J. B., 1909: *International Council for the Investigation of the Sea. Publications de Circonstance,* No. 44.
Fowle, F. E., 1918: *Smithson. Misc. Coll.,* 68, No. 8.
Hettner, A., 1918: *Ann. Phys., Leipzig,* 55.
Kincer. J. B., 1933: *Mon. Weath. Rev., Wash.,* 61.
Mossman, R. C., 1902: *Trans. Roy. Soc., Edin.,* 40.
Radcliffe Observatory, 1930: *Met. Obs.,* 55.
Rubens. H., and Aschkinass, R., 1898: *Ann. Phys. Chem.,* 64.
Schmidt, H., 1913: *Ann. Phys., Leipzig,* 42.
_____.1927 and 1934: "World weather records." *Smithson. Misc. Coll.,* 79 and 90.
Simpson, G. C., 1928: *Mem. R. Met. Soc.,* 3, No. 21.

# Carbon Dioxide Exchange Between Atmosphere and Ocean and the Question of an Increase of Atmospheric $CO_2$ during the Past Decades

*Roger Revelle and Hans E. Suess*

p. 39

Arrhenius, Svante, 1903: *Lehrbuch der kosmischen Physik* 2. Leipzig: Hirzel.

Bohr, C., 1899: Die Löslichkeit von Gasen in Flüssigkeiten. *Ann. d. Phys.* 68, p. 500.

Buch, K., 1933: Der Borsäuregehalt des Meerwassers und seine Bedeutung bei der Berechnung des Kohlensäuresystems. *Rapp. Cons. Explor. Mer.* 85, p. 71.

Callendar, G. S., 1938: The artificial production of carbon dioxide and its influence on temperature. *Quarterly Journ. Royal Meteorol. Soc.* 64, p. 223.

Callendar, G. S., 1940: Variations in the amount of carbon dioxide in different air currents. *Quarterly Journ. Royal Meteorol. Soc.* 66, p. 395.

Callendar, G. S., 1949: Can carbon dioxide influence climate? *Weather* 4, p. 310.

Chamberlin, T. C., 1899: An attempt to frame a working hypothesis of the cause of glacial periods on an atmospheric basis. *J. of Geology* 7, pp. 575, 667, 751.

Conway, E. J., 1942: Mean geochemical data in relation to oceanic evolution. *Proc. Roy. Irish Acad., B.* 48, p. 119.

Craig, H., 1953: The geochemistry of the stable carbon isotopes. *Geochim. et Cosmochim. Acta* 3, p. 53.

Craig, H., 1954: Carbon 13 in plants and the relationship between carbon 13 and carbon 14 variations in nature. *Journ. Geol.* 62, p.115.

Dingle, H. N., 1954: The carbon dioxide exchange between the North Atlantic Ocean and the atmosphere. *Tellus* 6, p. 342.

Erikson, E., and Welander, P., 1956: On a mathematical model of the carbon cycle in nature. *Tellus* 8, p. 155.

Fergusson, G. J., and Rafter, T. A.: *New Zealand C-14 Age Measurements III*. In press.

Fonselius, S., Koroleff, F., and Wärme, K., 1956: Carbon dioxide variations in the atmosphere. *Tellus* 8, p. 176.

Harvey, H. W., 1955: *The Chemistry and Fertility of Sea Water*.
Cambridge: University Press.

Hayes, F. N., Anderson, E. C, and Arnold, J. R., 1955: Liquid
scintillation counting of natural radiocarbon. *Proc. of the
International Conference on Peaceful Uses of Atomic Energy,
Geneva*, 14, p. 188.

Hutchinson, G. E., 1954: *In The Earth as a Planet*, G. Kuiper, ed.
Chicago: University Press. Chapter 8.

Munk, W., and Revelle, R., 1952: Sea level and the rotation of the earth.
*Am. Journ. Set.* 250, p. 829.

Nier, A. O., and Gulbransen, E. A., 1939: Variations in the relative
abundance of the carbon isotopes. *Journ. Am. Chem. Soc.* 61, p.
697.

Plass, G. N., 1956: Carbon dioxide theory of climatic change. *Tellus* 8,
p. 140.

Rafter, T. A., 1955: $C^{14}$ variations in nature and the effect on radiocarbon
dating. *New Zealand Journ. Sci. Tech. B.* 37, p. 20.

Rubey, W. W., 1951: Geologic history of sea water. *Bull. Geol. Soc.
Amer.* 62, p. 1111.

Slocum, Giles, 1955: Has the amount of carbon dioxide in the
atmosphere changed significantly since the beginning of the
twentieth century? *Monthly Weather Rev.* Oct., p. 225.

Suess, H. E., 1953: Natural Radiocarbon and the rate of exchange
of carbon dioxide between the atmosphere and the sea. *Nuclear
Processes in Geologic Settings*, National Academy of Sciences —
National Research Council Publication, pp. 52–56.

Suess, H. E., 1954: Natural radiocarbon measurements by acetylene
counting. *Science* 120, p. 5.

Suess, H. E., 1955: Radiocarbon concentration in modern wood. *Science*
122, p. 415.

Sverdrup, H. U., Johnson, M. W., and Fleming, R. H., 1942: *The
Oceans*. New York: Prentice-Hall, Inc.

United Nations, 1955: World requirements of energy, 1975-2000.
*International Conference on Peaceful Uses of Atomic Energy,
Geneva*, 1, p. 3.

Urey, H. C., 1952: *The Planets*. New Haven: Yale Univ. Press.

## Statement of Dr. James Hansen
*James Hansen*
p. 47

1. Hansen J., I. Fung, A. Lacis, D. Rind, G. Russell, S. Lebedeff, R. Ruedy and P. Stone, 1988, Global climate changes as forecast by the GISS 3-D model, *J. Geophys, Res.* (in press).
2. Hansen, J., A. Lacis, D. Rind, G. Russell, P. Stone, I. Fung, R. Ruedy and J. Lerner, 1984, Climate sensitivity: analysis of feedback mechanisms, *Geophys, Mono.*, 29, 130-163.
3. Manabe, S., R. T. Wetherald and R. J. Stauffer, 1981, Summer dryness due to an increase in atmospheric $CO_2$ concentration, *Climate Change*, 3, 347-386.
4. Manabe, S. and R. T. Wetherald, 1986, Reduction in summer soil wetness induced by an increase in atmospheric carbon dioxide, *Science*, 232, 626-628.
5. Manabe, S. and R. T. Wetherald, 1987, Large-scale changes of soil wetness induced by an increase in atmospheric carbon dioxide, *J. Atmos. Sci.*, 44, 1211-1235.
6. Hansen, J. and S. Lebedeff, 1987, Global trends of measured surface air temperature, *J. Geophys. Res.*, 92, 13, 345-13, 372; Hansen, J. and S. Lebedeff, 1988, Global surface air temperatures: update through 1987, *Geophys, Res. Lett.*, 15, 323-326.
7. Grotch, S., 1988, Regional intercomparisons of general circulation model predictions and historical climate data, Dept. of Energy Report, DOE/NBA-0084.

## The "Anthropocene"
*Paul J. Crutzen and Eugene F. Stoermer*
p. 69

1. Encyclopaedia Britannica, Micropaedia, IX, (1976).
2. G. P. Marsh, *The Earth as Modified by Human Action* (Cambridge, MA: Belknap Press, Harvard University Press, 1965).
3. W. C. Clark, in *Sustainable Development of the Biosphere*, W. C. Clark and R. E. Munn, eds. (Cambridge, U.K.: Cambridge University Press, 1986), chapter 1.
4. V. I. Vernadsky, *The Biosphere*, translated and annotated version from

the original of 1926 (New York: Copernicus, Springer, 1998).
5. B. L. Turner II et al., *The Earth as Transformed by Human Action* (Cambridge, U.K.: Cambridge University Press, 1990).
6. P. J. Crutzen and T. E. Graedel, in *Sustainable Development of the Biosphere*, chapter 9.
7. R. T. Watson, et al., in *Climate Change: The IPCC Scientific Assessment*, J. T. Houghton, G. J. Jenkins, and J. J. Ephraums, eds. (Cambridge, U.K.: Cambridge University Press, 1990), chapter 1.
8. P. M. Vitousek et al., *Science*, 277, 494, (1997).
9. E. O. Wilson, *The Diversity of Life* (New York: Penguin, 1992).
10. D. Pauly and V. Christensen, *Nature*, 374, 255-257 (1995).
11. E. F. Stoermer and J. P. Smol, eds., *The Diatoms: Applications for the Environmental and Earth Sciences* (Cambridge, U.K.: Cambridge University Press, 1999).
12. C. L. Schelske and E. F. Stoermer, *Science*, 173, (1971); D. Verschuren et al., *J. Great Lakes Res.*, 24 (1998).
13. M. S. V. Douglas, J. P. Smol, and W. Blake Jr., *Science* 266 (1994).
14. A. Berger and M.-F. Loutre, *C. R. Acad. Sci. Paris*, 323, II A, 1–16 (1996).
15. H. J. Schellnhuber, *Nature*, 402, C19-C23 (1999).

The Scientific Consensus on Climate Change
*Naomi Oreskes*
p. 75

1. A. C. Revkin, K. Q. Seelye , *New York Times* A1 (19 June 2003).
2. S. van den Hove, M. Le Menestrel, H.-C. de Bettignies, *Climate Policy* 2(1), 3 (2003).
3. See www.ipcc.ch/about/about.htm.
4. J. J. McCarthy, Ed. *Climate Change 2001: Impacts, Adaptation, and Vulnerability* (Cambridge Univ. Press, Cambridge, 2001), 21
5. National Academy of Sciences Committee on the Science of Climate Change, *Climate Change Science: An Analysis of Some Key Questions* (National Academy Press, Washington, D.C., 2001), 1
6. Ibid., 3
7. American Meteorological Society, *Bull. Am. Meteorol. Soc.* 84, 508 (2003).
8. American Geophysical Union, *Eos* 84(51), 574 (2003).

9. See www.ourplanet.com/aaas/pages/atmos02.html.
10. The first year for which the database consistently published abstracts was 1993. Some abstracts were deleted from our analysis because, although the authors had put "climate change" in their key words, the paper was not about climate change.
11. This essay is excerpted from the 2004 George Sarton Memorial Lecture, "Consensus in science: How do we know we're not wrong," presented at the AAAS meeting on 13 February 2004. I am grateful to AAAS and the History of Science Society for their support of this lectureship; to my research assistants S. Luis and G. Law; and to D. C. Agnew, K. Belitz, J. R. Fleming, M. T. Greene, H. Leifert, and R. C. J. Somerville for helpful discussions.

## Target Atmospheric CO$_2$: Where Should Humanity Aim?
*James Hansen et al.*
p. 81

1. Framework Convention on Climate Change, United Nations, 1992; www.unfccc.int.
2. Intergovernmental Panel on Climate Change (IPCC), *Climate Change 2007*, S. Solomon, Q. Dahe, M. Manning, et al. (eds.), (New York: Cambridge University Press, 2007), 996.
3. M. D. Mastrandrea, S. H. Schneider. Probabilistic Integrated Assessment of "Dangerous" Climate Change. *Science* (2004), 304: 571–575.
4. European Council, Climate Change Strategies. (2005), http://register.consilium.europa.eu/pdf/en/05/st07/st07242.en05.pdf
5. J. Hansen, M. Sato, Ruedy, et al. Dangerous Human-Made Interference with Climate: A GISS ModelE Study. *Atmos. Chem. Phys.* (2007), 7: 2287–2312.
6. J. Hansen, M. Sato., Greenhouse Gas Growth Rates. *Proc. Natl. Acad. Sci.* (2004), 101: 16109–16114.
7. J. Hansen, M. Sato, P. Kharecha, G. Russell, D. W. Lea, and M. Siddall, Climate Change and Trace Gases. *Phil. Trans. R. Soc. A.* (2007), 365: 1925–1954.
8. J. Hansen, L. Nazarenko, R. Ruedy, et al. Earth's Energy Imbalance: Confirmation and Implications. *Science* (2005), 308: 1431–1435.
9. L. D. D. Harvey, Dangerous Anthropogenic Interference, Dangerous

Climatic Change, and Harmful Climatic Change: Non-Trivial
Distinctions with Significant Policy Implications. *Clim. Change*
(2007), 82: 1–25.
10. H. D. Matthews, K. Caldeira. Stabilizing Climate Requires Near-
Zero Emissions. *Geophys. Res. Lett.* (2008), 35: L04705.
11. D. Archer, Fate of Fossil Fuel $CO_2$ in Geologic Time. *J. Geophys.
Res.* (2005), 110: C09S05.

### Causes of the Russian Heat Wave and Pakistani Floods
*Jeff Masters*
p. 89

L. Feudale and J. Shukla (2010), "Influence of Sea Surface Temperature
on the European Heat Wave of 2003 Summer. Part I: An
Observational Study," *Climate Dynamics*. DOI: 10.1007/s00382-
010-0788-0.
P. A. Stott, D. A. Stone, and M. R. Allen, "Human Contribution to the
European Heatwave of 2003," *Nature* (2 December 2004), 432:
610–614. DOI:10.1038/nature03089. (A free version of the paper,
presented at a conference, is available at www-atm.physics.ox.ac.
uk/main/Science/posters2005/2005ds.pdf.)
The World Meteorological Organization (WMO) has posted an analysis of
the recent extreme weather events, concluding, "The sequence of current
events matches IPCC projections of more frequent and more intense
extreme weather events due to global warming." (See www.wmo.
int/pages/mediacentre/news/extremeweathersequence_en.html.)

### from *The Green Collar Economy: How One Solution Can Fix Our Two Biggest Problems*
*Van Jones*
p. 211

1. Grover Norquist on National Public Radio, May 25, 2001, http://
www.npr.org/templates/story/story.php?storyId=1123439.
2. Bruce Springsteen on the Vote for Change Tour, October 10, 2004.
3. U.S. Department of Labor Bureau of Statistics, "Employment
Situation Summary," May 2008, http://www.bls.gov/news.release/
empsit.nr0.htm.

4. Bracken Hendricks and Jay Inslee, *Apollo's Fire* (Washington, DC: Island Press, 2008).
5. Carla Din, "Finding Opportunity in Crisis," Yes Magazine (Fall 2004), http://www.yesmagazine.org/article.asp?ID=1030.
6. Interview with Elsa Barboza, February 2008.
7. Elsa Barboza, "Organizing for Green Industries in Los Angeles," Race, Poverty and the Environment 13, no. 1 (Summer 2006), http://www.urbanhabitat.org/node/525.
8. In June 2007, the city council of LA established a City Retrofit Jobs Task Force made up of Apollo Alliance representatives, council members, and employees of various City agencies. The task force is identifying the workforce needs, potential job-training providers, and funding sources.
9. Joanna Lee, Angela Bowden, and Jennifer Ito, Green Cities, Green Jobs, May 2007, http://www.greenforall.org/resources/green-cities-gree-jobs-by-joanna-lee-angela/download.
10. Lee, Bowden, and Ito, Green Cities, Green Jobs.
11. Lee, Bowden, and Ito, Green Cities, Green Jobs.
12. Center on Wisconsin Strategy (COWS), http://www.cows.org/collab_projects_detail.asp?id=54.

**Climate Generation: The Evolution of the Energy Action Coalition's Strategy**
*Billy Parish*
p. 225

1. http://www.cookpolitical.com/
2. http://itsgettinghotinhere.org/2010/01/12/climate-generation-in-2010-go-big/
3. http://www.grist.org/article/copenhagen-getting-past-the-urgency-trap/
4. http://itsgettinghotinhere.files.wordpress.com/2010/01/eac-doa-history.pdf
5. http://www.presidentsclimatecommitment.org/signatories/list
6. http://stepitup2007.org/

## This Is Fucked Up—It's Time to Get Mad, and Then Busy
*Bill McKibben*
p. 251

1. http://www.democracynow.org/2010/7/29/headlines/2000_2009_marked_warmest_decade_on_record
2. http://www.scientificamerican.com/article.cfm?id=phytoplankton-population
3. http://www.wunderground.com/blog/JeffMasters/comment.html?entrynum=1546
4. http://www.wunderground.com/blog/JeffMasters/comment.html?entrynum=1498&tstamp=
5. http://www.nytimes.com/2010/07/28/opinion/28friedman.html
6. http://www.rollingstone.com/politics/news/12697/64796
7. http://thebreakthrough.org/ideas.shtml
8. http://putsolaron.it/
9. http://www.commondreams.org/views/121000-104.htm
10. http://www.tomdispatch.com/post/174949/mike_davis_welcome_to_the_next_epoch
11. http://www.nytimes.com/2010/07/25/opinion/25friedman.html
12. http://www.350.org/
13. http://www.foreignpolicy.com/articles/2009/11/30/the_fp_top_100_global_thinkers?page=full
14. http://arxiv.org/pdf/0804.1126
15. http://www.flickr.com/photos/350org/sets/
16. http://www.350.org/
17. http://appalachiarising.org/
18. http://www.peacefuluprising.org/
19. http://www.care2.com/causes/trailblazers/blog/the-greatest-generation
20. http://interfaithpowerandlight.org/
21. http://energyactioncoalition.org/

## The Population Myth
*George Monbiot*
p. 269

1. Optimum Population Trust, Gaia Scientist to be OPT Patron (August 26, 2009). See www.optimumpopulation.org/releases/opt.release26Aug09.htm

2. David Satterthwaite, The Implications of Population Growth and Urbanization for Climate Change, Environment & Urbanization (September 2009) Vol 21(2) : 545–567. DOI: 10.1177/0956247809344361.

3. See www.foei.org/en/publications/pdfs-members/economic-justice/gasnigeria.pdf.

4. For example, Satterthwaite cites the study by Gerald Leach and Robin Mearns, Beyond the Woodfuel Crisis: People, Land, and Trees in Africa, (London: Earthscan Publications, 1989).

5. See www.ybw.com/auto/newsdesk/20090802125307syb.html.

6. See www.jameslist.com/advert/5480.

7. See http://machinedesign.com/article/118-wallypower-a-high-end-power-boat-0616.

8. 15 US gallons per nautical mile = 56.775 liters per nautical miles = 31 liters per kilometer.

9. John Harlow, Billionaire Club in Bid to Curb Overpopulation, the Sunday Times (May 24, 2009).

10. Wolfgang Lutz, Warren Sanderson, and Sergei Scherbov, The Coming Acceleration of Global Population Aging. Nature (January 20, 2008). DOI:10.1038/nature06516.

11. U.N. Department of Economic and Social Affairs, World Population Prospects: The 2004 Revision (2005). See www.un.org/esa/population/publications/sixbillion/sixbilpart1.pdf.

## Global Warming Twenty Years Later: Tipping Points Near
James Hansen
p. 275

1. J. Hansen, M. Sato, P. Kharecha, D. Beerling, R. Berner, V. Masson-Delmotte, M. Raymo, D. L. Royer, J. C. Zachos, "Target Atmospheric CO2: Where Should Humanity Aim?" Available at http://arxiv.org/abs/0804.1126 and http://arxiv.org/abs/0804.1135.

2. The proposed "Tax and 100 percent dividend" is based largely on the cap-and-dividend approach described by Peter Barnes in Who Owns the Sky: Our Common Assets and the Future of Capitalism (Washington, D.C.: Island Press, 2001).

**Climate Change: An Evangelical Call to Action**
*The Evangelical Climate Initiative*
p. 309

1. Cf. "For the Health of the Nation: An Evangelical Call to Civic Responsibility," approved by National Association of Evangelicals, October 8, 2004.
2. Intergovernmental Panel on Climate Change 2001, Summary for Policymakers; www.grida.no/climate/ipcc tar/wg1/007.htm. (See also the main IPCC website, www.ipcc.ch.) For the confirmation of the IPCC's findings from the U.S. National Academy of Sciences, see "Climate Change Science: An Analysis of Some Key Questions" (2001), http://books.nap.edu/html/climatechange/summary.html. For the statement by the G8 Academies (plus those of Brazil, India, and China) see "Joint Science Academies Statement: Global Response to Climate Change" (June 2005), http://nationalacademies. org/onpi/06072005.pdf. Another major international report that confirms the IPCC's conclusions comes from the Arctic Climate Impact Assessment. See their "Impacts of a Warming Climate" (Cambridge University Press, November 2004), p.2, http://amap.no/ acia/. Another important statement is from the American Geophysical Union, "Human Impacts on Climate" (December 2003), www. agu.org/sci soc/policy/climate change position.html. For the Bush administration's perspective, see "Our Changing Planet: The U.S. Climate Change Science Program for Fiscal Years 2004 and 2005," p.47, www.usgcrp.gov/usgcrp/Library/ocp2004-5/default.htm. For the 2005 G8 statement, see www.number-10.gov.uk/output/Page7881.asp.

# Index

energy demand, 150, 159
  total energy consumption, 148
chlorofluorocarbons, 56, 71, 99–102, 123,
  166
Christy, John, 173, 179, 187
Chu, Steven, U.S. Energy Secretary, 239
Cinergy, 314
Clean Air Act, 98
Clearinghouse in Environmental Advocacy
  and Research, 115
climate debt, 240–247
Climate Generation series, 225
Clinton, Bill, 127, 131, 160
coal
  as resource in China, 142
  production of, 23
coccolithophores, 392–393
Columbia University, 275, 386, 395
Copenhagen
  climate change summit, 239–240
coral reefs, 385–386
  bleaching of, 390
Corker, Bob, 14
Correa, Rafael, 244
Costa Rica
  proliferation of the Aedes aegyptii
  mosquito, 110
Crichton, Michael, 192, 193
Croll, James
  theory of climate change, 20, 29–30
Crutzen, Paul, 68, 69
Cyprus Minerals
  antienvironmentalist agenda, 115

**D**

DeChristopher, Tim, 257
Declaration of Independence from Dirty
  Energy, 227
De Marchi, Luigi, 28–29
dengue fever, 12, 110
Deng Xiaoping, 156
Doolittle, John, 124
Draper, Lynda, 100–101
Duke Energy, 254, 314
DuPont, 314
Dyson, Freeman, 190

**E**

*Earth Odyssey* (book), 134–135
*End of Nature, The* (book), 7, 11, 68, 292,
  293. *See* McKibben, Bill (author)
Energy Action Coalition, 224–226, 230
  evolution of, 225
Ensley, Oreatha, 219
Environmental Defense Fund, 250, 252
*Environmental Values in American Culture*
  (book), 113
*Environment Writer,* 116
Essex, Chris, 190
European Union Commission, 123
ExxonMobil, 13, 122, 254, 255, 279

**F**

Fabry, Victoria, 377, 396
Foil Wars, 355
Fossil Fools Day (April 1, 2004), 226
fossil fuel
  combustion of, 40
Fox, C.J., 34, 35
Friedman, Thomas (Tom), 253, 255
Friends of the Earth, 257
Fusheng, Wei, 155

**G**

Gelbspan, Ross, 104, 105, 164
General Electric, 100, 314
George C. Marshall Institute, 189
GHGs. *See* greenhouse gases
Gleick, Peter, 339
Global Business Network, 324
Global Climate Coalition, 111
global warming, 54, 113, 121
  and Christianity, 309
  and U.S. national security, 319–348
Gore, Al, 74, 96, 97, 104, 126, 127, 208,
  209, 263
Great Barrier Reef, 385
Great Leap Forward campaign, 137
Great Russian Heat Wave of 2010, 89
Green-Free Month, 233
greenhouse effect, 32, 46–47, 49–53, 109,
  124, 152, 297
  as relates to global warming, 49

# Permissions

Sally Bingham, Sermon on John 5:1-9, originally delivered at Stanford Chapel, May 9, 2010. Reprinted by permission of Sally Bingham.

Adrienne Maree Brown, "The Green Generation," originally published in *Wiretap* magazine, November 16, 2007, http://www.wiretapmag.org/environment/43304/. Reprinted by permission of Adrienne Maree Brown.

Michael Crichton, excerpt from *State of Fear* (New York: HarperCollins, 2004), pp. 39–51. Copyright © 2004 by Michael Crichton. Reprinted by permission of HarperCollins.

Paul J. Crutzen and Eugene F. Stoermer, "The Anthropocene." Originally published in the *International Geosphere-Biosphere Programme Newsletter*, no. 41 (May, 2000). Reprinted by permission of the International Geosphere-Biosphere Programme.

Ross Gelbspan, excerpt from *The Heat Is On* (New York: Basic Books, 1998), pp. 33–49. Copyright © 1998 by Ross Gelbspan. Reprinted by permission of Perseus Books Group.

Al Gore, excerpt from *Earth in the Balance* (New York: Houghton Mifflin Harcourt, 2000), pp. 292–295. Copyright © 2000 by Al Gore. Reprinted by permission of Houghton Mifflin Harcourt.

James Hansen, "Target Atmospheric $CO_2$: Where Should Humanity Aim?" originally published in the *Open Atmospheric Science Journal*, Vol. 2, pp. 217-231, and "Global Warming Twenty Years Later: Tipping Points Near," www.columbia.edu/~jeh1/2008/TwentyYearsLater_20080623.pdf. Reprinted by permission of James E. Hansen.

Mark Hertsgaard, Excerpt from *Earth Odyssey* (New York: Broadway Books, 1998), pp. 156–170, 181–188. Copyright © 1998 by Mark Hertsgaard. Reprinted by permission of Broad-way Books, a division of Random House, Inc.

Van Jones, excerpt from *The Green Collar Economy* (New York: HarperOne, 2008), pp. 108–119. Copyright © 2008 by Van Jones. Reprinted by permission of HarperCollins.

Naomi Klein, "Climate Rage," originally published in *Rolling Stone*, November 11, 2009, http://www.naomiklein.org/articles/2009/11/climate-rage. Reprinted by permission of Naomi Klein.

Elizabeth Kolbert, "The Darkening Sea." Originally published in *The New Yorker*, November 20, 2006, pp. 66–75. Reprinted by permission of Elizabeth Kolbert.

Jeff Masters, "Causes of the Russian Heat Wave and Pakistani Floods," originally published in *Weather Underground*, http://www.wunderground.com/blog/JeffMasters/comment.html?entrynum=1576. Reprinted by permission of Jeff Masters.

Bill McKibben, "This Is Fucked Up—It's Time to Get Mad, and then Busy," originally published in *The Huffington Post*, August 4, 2010, http://www.huffingtonpost.com/bill-mckibben/this-is-f-cked-up----its_b_670347.html. Reprinted by permission of Bill McKibben.

Bill McKibben, excerpt from *The End of Nature* (New York: Anchor Books, 1999), pp. 86–91. Copyright © 1989 and renewed © 1999 by Bill McKibben. Reprinted by permission of Random House, Inc.

Naomi Oreskes, "The Scientific Consensus on Climate Change," originally published in *Science*, December 3, 2004: Vol. 306 no. 5702 p. 1686. Reprinted by permission of The American Association for the Advancement of Science. Copyright © 2004, American Association for the Advancement of Science.

Billy Parish, "Climate Generation: The Evolution of The Energy Action Coalition's Strategy," originally published in *It's Getting Hot In Here*, http://itsgettinghotinhere.org/2010/01/13/climate-generation-the-evolution-of-the-energy-action-coalitions-strategy/. Reprinted by permission of Billy Parish.

Roger Revelle and Hans E. Suess, excerpt from "Carbon Dioxide Exchange Between Atmosphere and Ocean and the Question of an Increase of Atmospheric $CO_2$ during the Past Decades." Originally published in *Tellus*, Vol. 9, Issue 1, pp. 18–27, February 1957, by Wiley-Blackwell. Reprinted by permission of Wiley-Blackwell.

Arundhati Roy, "The Briefing," excerpt from *Field Notes on Democracy: Listening to Grasshoppers* (Chicago: Haymarket Books, 2009), pp. 202–210. Copyright © 2009 by Arundhati Roy. Reprinted by permission of Haymarket Books.

Vandana Shiva, "Climate Change and Agriculture," originally published February 24, 2011, http://www.vandanashiva.org. Reprinted by permission of Vandana Shiva.

Mike Tidwell, "To Really Save the Planet, Stop Going Green," originally published in *The Washington Post*, December 6, 2009, http://www.washingtonpost.com/wp-dyn/content/article/2009/12/04/AR2009120402605.html. Reprinted by permission of Mike Tidwell.

John Vidal, "Nepal's Farmers on the Front Line of Global Climate Change," originally published in *The Guardian*, December 2, 2006, http://www.guardian.co.uk/environment/2006/dec/02/christmasappeal2006.frontpagenews. Copyright © Guardian News & Media Ltd 2006. Reprinted by permission.

# Image Permissions